THIS IS NOT
A WEASEL

THIS IS NOT
A WEASEL

A CLOSE LOOK AT NATURE'S MOST
CONFUSING TERMS

Philip B. Mortenson

WILEY

John Wiley & Sons, Inc.

Published by John Wiley & Sons, Inc., Hoboken, New Jersey
Published simultaneously in Canada

Design and production by Navta Associates, Inc.

For general information about our other products and services, please contact our Customer Care Department within the United States at (800) 762-2974, outside the United States at (317) 572-3993 or fax (317) 572-4002.

Wiley also publishes its books in a variety of electronic formats. Some content that appears in print may not be available in electronic books. For more information about Wiley products, visit our web site at www.wiley.com.

Library of Congress Cataloging-in-Publication Data:

Mortenson, Philip.
   This is not a weasel : a close look at nature's most confusing terms / Philip Mortenson.
      p. cm.
   ISBN 0-471-27396-1 (Cloth)
   1. Biology—Nomenclature.   I. Title.
   QH83 .M69   2003
   570'.1'4—dc21   2003005318

Printed in the United States of America

10 9 8 7 6 5 4 3 2 1

For Joseph, Kieran, Erin,
and all future grandchildren

Contents

Acknowledgments

As with all literary projects, none are ever conceived and brought to fruition in a vacuum. The process of conceiving, researching, writing, and producing this book was no different. Among those to whom I owe a good deal of gratitude are my wife JoAnn, who was an ever-constant source of support and encouragement, from the inception of the idea for the book right through to its publication. My children, Kelly, Ryan, and Adrienne, were supportive and helpful in their many suggestions and criticisms. My thanks, too, to my agent, Marlene Connor Lynch, who guided the project with invaluable advice and perseverance. At the production end, I must acknowledge the thoughtful counsel and gentle handling of the manuscript by my editor at John Wiley and Sons, Jeff Golick, who not only took a chance on an unknown but did so with patient respect. I also thank my production editor, Hope Breeman, for her kind and competent guidance of the book through its production, and my copyeditor, Patricia Waldygo, for her meticulous work on the manuscript.

INTRODUCTION

T his book examines nature's most confusing terms: pairs or groups of words that often trip us up because we don't quite understand how they differ. It embraces all facets of life and addresses these troublesome terms by delving into the nitty-gritty of their critical differences.

What kinds of troublesome terms? Well, take the popular vegetable *Ipomoea batatas,* which many people call a sweet potato, and others call a yam. Although both terms have been long regarded as synonyms, some people feel this is a grave mistake. If we concede that they may be right, the question then is: which term, "sweet potato" or "yam," is correct? A can of *Ipomoea batatas* taken from the grocer's shelf would appear to settle the question; its label reads: "*Princella Sweet Potatoes.* Cut Yams in Light Syrup." So there we have it: this enduring staple of Thanksgiving dinners may rightly be called either a "sweet potato" or a "yam."

Wrong! This tasty, oblong root is no more a yam than a hazelnut is an apple. In fact, we should be thankful that true yams never make it to our dinner table; they are disagreeable tubers whose bitter taste can sometimes be removed only by boiling.

The Betty Crockers of the world, having been misled by grocers, product labels, garden centers, and ill-informed cookbook authors, may now be scratching their heads. If these vegetables are sweet potatoes, then what are we doing calling them yams? And what is the difference between a yam and a sweet potato anyway?

The answers, of course, lie within these pages, as do clues for unraveling a host of nature's other confusing terms. They not only fully discuss each subject but often make ample use of examples, cite conflicting opinions, debunk common misconceptions, and, where necessary, expose the flagrant misuse of terms.

Obviously, it is not nature's fault but our own that confusion has arisen over her many facets. Most of us recognize that "dolphin" and "porpoise" refer to two distinct animals, but few of us know the specifics that define each animal. What makes a dolphin a dolphin, a porpoise a porpoise, and what makes them different from each other? Other terms tend to be confusing because we are simply unsure of them. Consider "buzzard" and "vulture": two different birds or simply different names for the same bird? Then there are outright misnomers. Although the Barbary macaque is an Old World monkey, its lack of a tail makes it easy to mistake for an ape, which is why we readily accept its more familiar name: Barbary ape.

Even more confusing when authorities contradict each other although, to be fair, this usually stems from a legitimate effort to better understand and describe our world. By its very nature, the pursuit of knowledge continually generates conflicting opinions and conclusions. This is why, although we may have our own ideas about the differences between mushrooms and morels, mycologists (those who study such things) have yet to reach a consensus on the subject.

Then there are instances in which confusion arises because people of different cultures or societies define terms differently. In the United States, for example, the word "corn" refers to only one kind of crop, whereas in Europe the term may denote anything from barley to beans. Confusion can also occur when a popular understanding of a term collides with its scientific meaning. Just because we call a string bean a vegetable doesn't mean that it isn't also a fruit, which raises the question: if that's true, then just what do the terms "fruit" and "vegetable" really mean?

But the worst culprits are authoritative sources that blindside us with erroneous information. Dictionaries, encyclopedias, textbooks, science programs, news reports, and popular science publications are all guilty of occasionally misstating the facts. One source mistakenly contends that a sardine is no more than a herring packed in a sardine can. Another states that cicadas and aphids are arachnids—they are, in fact, insects. Yet another source falsely assigns the bulrush, a member of the sedge family, to the rush family. And one well-known nature show on television stated that the Kodiak bear is a grizzly; it is not. These blunders are unfortunate because the last thing we need when sorting out nature's mysteries is more confusion, especially when it comes from sources on which we have learned to rely.

To resolve the difference within each set of confusing terms, I have used several approaches, each guided by the subject matter and the degree of confusion that surrounds it. In cases not involving significant differences of opinion, shifts in definition, or ambiguous meanings, the discussions are fairly straightforward. These discussions require little more than a brief assessment of specific characteristics. In other cases, as in the centipede–millipede issue, not only are the terms examined, but also comparative lists help to illustrate their differences. In others, such as the turtle, tortoise, and terrapin discussion, a history of the three terms is an important element. And where the subject is somewhat complex, as with Kingdom, Phylum, Class, and so on, several forms of assessment are used.

But whatever the approach, my goal has been to present scientific information and definitions wherever possible, rather than to focus on popular meanings. Unfortunately, this has not been without its problems. As indicated, scientists often have differing ideas about what a term does and does not mean. Where these opposing views are significant, I have tried to give them due recognition. Yet sometimes science has little to say on a subject, in which case a consensus of popular usage is presented, along with any reasonable alternative views.

In all, the reader will find 243 terms assembled into 78 discussions. And whereas some subjects are quite obvious, others may at first seem a bit odd. But whatever the case, each was crafted to provide the most useful answer possible. Unfortunately, this often meant employing unfamiliar scientific terms, as well as taxonomic concepts. So, wherever an unfamiliar term is introduced, it is followed by a brief definition. Taxonomic names, while often considered an annoyance, are used extensively throughout the book because in many cases there is simply no better way to represent the different forms of a plant and an animal, and, also, taxonomic names are indispensable in student research. Yet I must caution that although the taxonomic classifications and names presented are current, they may not be the only ones in use, nor are they exempt from change. I am sure that as you read this, changes in classification have already occurred. In any case, I strongly recommend to the reader the two opening discussions on nomenclature and taxonomy.

The reader may also notice the profusion of "weasel" words, such as "but," "sometimes," "however," and "except." The extensive use of these words arose because, unlike the customary practice of presenting information as well-accepted fact, I feel that a more realistic and helpful approach is to convey the uncertainty that often accompanies a term (a common affliction in science). Because the terms that comprise each discussion often have diverse origins (science, popular usage, folk customs, foreign languages, and the like), have undergone significant changes in meaning, or are embroiled in ongoing disputes, precise distinctions are not always possible. Therefore, some conclusions may be disappointing—or maybe the ambiguity will be welcomed. But even where distinctions are well established, it must be remembered that they are all subject to change. Little is written in stone, even, or especially, in science.

1

NOMENCLATURE AND TAXONOMY

Nomenclature (Scientific ● Common)

When casually discussing nature, whether verbally or in print, we typically use pretty informal language. Simple, everyday terms for plants and animals are more than adequate for expressing ourselves. We do not talk about *Malus* and *Citrus sinensis* (taxonomic terms) when referring to apples and oranges because it is simply not effective communication. However, we do use such terms when we want to make specific scientific distinctions—say, between the prairie crab apple, which botanists identify as *Malus ioensis,* and the sweet crab apple, which they have designated *Malus coronaria.*

Because this book deals with differences, many of them technical and somewhat detailed, the reader will encounter quite a few such scientific names, some of which may appear not only needlessly elaborate but also virtually unpronounceable—although nothing as tongue twisting as *Paracoccidioides brasilensis,* a lung-infecting fungus, or *Brachyta interrogationis interrogationis* var. *nigrohumeralisscutellohumeroconjuncta,* a longhorn beetle. As for pronunciation, unless one really feels that the word will be used in conversation, there is no need to be concerned with "getting it right."

But why does science eschew common names and instead encumber itself with such a seemingly inexplicable vocabulary? Because, in part, "common" does not necessarily mean "universal." What may be an organism's common name in one part of the country may be quite different in another, making any initial exchange of information suspect. For example, in the Midwest the thirteen-lined ground squirrel *(Spermophilus tridecemlineatus)* is commonly, but mistakenly, called a gopher, which is really the rightful name of a much larger rodent *(Geomys bursarius)*. Furthermore, some organisms seem to have acquired a different name every time they were spotted in a new locale, such as the mullein, a plant of the figwort family that is saddled with almost 150 aliases.

Unfortunately, this is just the sort of predicament that faced the entire biological science community up until the mid-eighteenth century: scientists from different countries and even individual scientists within the same country sometimes used different names for the same plants and animals. Early attempts to rectify the predicament by using descriptive Latin words to identify each species only led to new problems. Species often ended up bearing unwieldy labels, such as the red-winged blackbird's *Sturnus niger alis superne rubentibus* and the catnip plant's *Nepeta floribus interrupte spicatus pedunculatis*. Ultimately, it was the Swedish botanist Carolus Linnaeus[1] who remedied the problem by establishing a specific methodology for naming all organisms. He devised a binomial nomenclature that gave each species its own, and purposely unique, two-part name.

In accordance with this system, scientific names of species (viruses excepted) always consist of at least two terms—for example *Homo sapiens*. The first term *(Homo)* is the genus (plural, genera) it belongs to, and the second term *(sapiens)*, called the specific epithet, its species group. An organism is never referred to by its specific term alone: both terms must be used or indicated. In print, genus names always begin with a capital letter and the species name with a lowercase letter, and both are always italicized. If all the species of a genus are included in a statement, scientists will often indicate only its genus, followed by "spp." If the genus is in doubt, a question mark will follow the term. This most often occurs in classifying fossils or when a family is undergoing reorganization.

Whenever a scientific name is first mentioned in a piece of literature, both terms are used. Any subsequent use of the name in that work may be abbreviated by using the initial letter of the genus plus the full species name—for example, *H. sapiens*. Other species of the same genus that follow may be likewise abbreviated, but in all cases the references must be unambiguous. For subspecies, after the genus and species names have been established, these may be indicated by their initial letters, with the subspecies name written in full.

While these rules probably appear quite reasonable, the question may have occurred: why not use more familiar words? Instead of *Geomys bursarius*, why didn't scientists just label the animal "*plains gopher*" or some other common name, using simple everyday terms? In answer: the foreign ring to these names is a result of using Latin or Latinized words and word roots, the language of science that was favored by European scholars at the time binomial nomenclature was introduced (in some cases, Greek is also used but in a Latin form).

As a dead language, Latin is ideal because it is free of any nationalistic conceit and bias—averting any potential squabbling over which language has the best common name for an organism. It is also functional, in that suffixes are easily appended, forming terms that carry noteworthy information.[2] For instance, the scientific name for the Eastern Hog-Nosed skunk is *Conepatus leuconotus,* which comes from four descriptive terms: *konis,* "dust"; *pateo,* "I walk"; *leukes,* "white"; and *notos,* "the back." Linked together, they tell us that the animal is a "white back that roams the open or desert-like country." Unfortunately, scientific names do not always accurately describe an organism, which, for those who are familiar with Latin, may be a bit misleading. The pronghorn's genus name, for example, is *Antilocapra,* a name derived from *anthlops* (for antelope) and *capra* (for she goat); the pronghorn is neither. But scientific names are not always so formal or rooted in the precepts of academia. Naming is a privilege usually reserved by the first scientist to describe an organism and, as such, may reflect a personal interest, rather than some particular character of the organism. One scientist named two new species of wasps *Polemistus chewbacca* and *P. yoda* after characters in the movie *Star Wars.* And Godzilla, the famous Japanese movie monster, was the source of the crustacean genus name *Godzillius.*

As for common names, Robert Bakker, one of the world's foremost authorities on dinosaurs, summed up the latitude scientists enjoy in naming their discoveries. Bakker observed, "[There are] no rule books to name dinosaurs. It can be Urdu. It can be Hindu. It can be an anagram [although if it is also to be a taxonomic name it must be pronounceable]. It could be nonsense syllables—as long as they are not overtly insulting to some colleagues."[3] And as if to underscore the point, Bakker named one of his recent dinosaur finds Big Ed. However, under the codes governing the spelling of zoological organisms, if "Big Ed" were to be a taxonomic name, it would at least have to be Latinized with a Latin termination.

Yet as clever as these names may be, most scientific names are not concocted willy-nilly. The creation of almost all such names is guided by various international codes, each code determined by the discipline it serves. Formal terms involving animals are governed by the rules of the *International Code of Zoological Nomenclature,* those of wild plants by the *International Code of Botanical Nomenclature,* cultivated plants by the *International Code of Nomenclature for Cultivated Plants,* and bacteria by the *International Code of Nomenclature of Bacteria.* And viruses, although not regarded as living organisms, are nonetheless also named and classified (in an order with families, subfamilies, etc.) according to the *International Code of Virus Classification and Nomenclature*. These international codes do not regulate all ranks, just the more specific, lower ones. The zoological code, for instance, concerns itself only with those names of taxa in the ranks of superfamily down through subspecies.

While all codes follow taxonomic and Latin grammar rules, they are independent as to the terms they allow to be used. Therefore, the same term may be used in different fields of science. *Corydalis,* for example, is used in botany for a genus of herbs and in entomology for a genus of predatory insects called Dobsonflies. Likewise, *Ricinus* is the genus name for both the castor bean and the blood-sucking bird louse. Such duplication may also occur among the unregulated ranks within a discipline. In zoology "Decapoda" is used to designate not only one of its crustacean orders but one of its mollusk suborders as well. Organisms also have had different designations in separate disciplines. Until just recently, the troublesome plant/animal-like euglena could be found listed in the algal family Euglenaceae by botanists and in the protozoan family Euglenidae by microbiologists.

This illustrates the fact that scientific names, as well thought out as they may be, are not engraved in stone. While a change in name is not at all uncommon, it is never done casually but rather because of some compelling rationale. The name may already be in use or an older name may be found to take precedence, but most commonly, it is because the organism is reclassified. The echidna, or spiny anteater, is a perfect example.

When first discovered and described in 1792, the echidna was thought to be a relative of the anteater of South America and was therefore classified in the anteater genus under the name *Myrmecophaga aculeta* (prickled anteater). However, on closer examination in 1802, the echidna was found to be not a placental animal like the anteater but a monotreme like the platypus and was therefore reassigned to the platypus genus *Ornithorhynchus* (birdlike snout), with the species label of *hystrix* (porcupine-like). But shortly thereafter, scientists decided that this strange animal deserved its own genus, so they renamed it *Echidna hystrix* after the Greek goddess Ekhidna, who was half reptile and half mammal. Then in 1876 a new form of echidna was discovered in New Guinea, one with a much longer snout and shorter spines. These and other features made it so different from *E. hystrix* that it clearly required its own genus. Faced with creating a new genus name, zoologists decided that some shared feature should unite the names of both echidna genera. This feature would be their tongues. So *E. hystrix* was renamed *Tachyglossus* (rapid tongue) *aculeatus* (with prickles) and the New Guinea long-snouted echidna received the name *Zaglossus* (great tongue) *bruijni* (from the name of a noted naturalist).

Although such reclassifications will continue, they will probably never reach the extent they did during the trend away from the "splitting" approach to classification to a "lumping" approach. The splitters, who held sway into the early twentieth century, often regarded an organism's unique characteristics as compelling reasons to separate it, taxonomically, from similar organisms. This typically resulted in genera with a seeming overabundance of species (see "Brown Bear," chapter 7). But as biologists reevaluated their approach to taxonomy, they began to see the wisdom in lumping together those species with close similarities, which effectively reduced the number of genera and species. The net result was that many organisms lost their old designations, sometimes receiving not only a new species label but a new genus label as well.

One of the basic tools in classifying an organism is the type of words used. When one forms a binomial name, the genus term is always a singular noun (e.g., "tongue"), whereas the species name may be a noun, an adjective, or a Latin rendering of a proper name. Except for the kingdoms, which need only be a plural noun, the taxa (taxa [singular, taxon] are specific groups within a rank) of other ranks are formed from plural adjectives used as nouns and are often appended with specific endings. These endings often serve as an immediate indicator of the rank of the taxon. For example, within the ranking of the koala bear are the taxa Phalangeria and Phalangeridae. The "ia" ending indicates that this is its order name, and the "idae" indicates its family name. Listed as follows are taxa endings for the plant and animal kingdoms. The parenthetical endings after "varies" are only typical; often they are found in only one taxon of a higher rank, and may not be the only such ending in that rank.

	Taxa Endings	
Rank	**Plants**	**Animals**
Phylum/Division	-phyta	varies (-a)
Subphylum/Division	varies (-phytina)[4]	varies (-data)
Class	varies (-opsida)[4]	varies (-lia)
Subclass	varies (-idae)[4]	varies (-ia)
Superorder[5]	varies (-anae)	-orpha
Order	-ales	varies (-formes)
Suborder	-ineae	varies (-morpha)
Superfamily[5]	N/A	-oidea

continued on next page

continued from previous page

	Taxa Endings	
Rank	**Plants**	**Animals**
Family	-aceae	-idae
Subfamily	-oideae	-inae
Tribe	-eae	-ini

Kingdom ● Phylum ● Class ● Order ● Family ● Genus ● Species

These seven terms represent one of mankind's most ambitious undertakings: the continuing effort to uncover and describe the relationships that unite Earth's incredible variety of life. In addressing this task, two scientific disciplines are particularly crucial, systematics and its subdivision, taxonomy. Systematics deals with the kinds, diversity, and evolutionary history of organisms, which enables taxonomy to identify, name, and classify them into related groups. Together, the two work to assign every living thing—past and present—a position on the tree of life. (Often both disciplines are collectively indicated by either term.) No simple task, this immense undertaking assesses a variety of factors that ranges from ancestry to physical characteristics to gene sequencing.

The working assumption behind this endeavor is that billions of years ago, all life was born of a common ancestry and that every subsequent form of life, no matter how unique, is related to all others. Under this principle, the more temporally close an organism is linked to another in certain shared characteristics, the more closely related they are apt to be. We, as *Homo sapiens,* are more closely related to the gorilla than to the sea squirt, but not as closely as we are to our 500,000-year-old relative *Homo erectus.* An incalculable number of evolutionary links, many unknown and undiscoverable, connects these four organisms, but the clues pointing to their evolutionary kinship are unequivocal. This does not mean that certain lower forms of life such as sea squirts were necessarily our ancestors (sea squirts, in fact, appear to be an evolutionary dead end), but only that their appearance on Earth arose from a different branch of the same primeval stock.

In their work, taxonomists classify life onto a hierarchical framework consisting of increasingly restrictive categories, much in the same way one could rank a deck of playing cards. To illustrate: if we take an entire deck of cards as the broadest division, all cards would be members. Each also belongs to a smaller group of either red or black cards. By themselves, the red cards easily fall into one of two other groups: hearts or diamonds. Considering just one suit, we can make a distinction between the court cards and the numbered cards, giving us two yet smaller groupings. Taking the 10 numbered cards, we can further divide them into two groups of odd- and even-numbered members. From this point, we have the individual cards themselves.

Using such a hierarchical system of classification, we can pick any card from the deck and assign it a place on the tree of playing cards. If our lone card happens to be the three of diamonds, we know that it is a member of an odd card group, which is a member of the noncourt group, which in this case is a member of the diamond group, which is a member of the red group, which is a member of the group of all cards. This is a tree of five ranks, from the rank of deck to the rank of individual card. An important point to keep in mind in this analogy is that there is no inherent connection between the various cards. A card manufacturer could make all the clubs red and all the hearts black and not change the fundamental nature of the deck.

The beauty of this system is that at a glance, one can determine how closely individual cards are "related" to one another (in this particular scheme). The three of clubs may seem very much akin to our three of diamonds—being of equal value and rank (the fifth)—yet it is apparent that they show no common connection until we move all the way up to the first rank. That they are both "threes" is trivial, in this case. Had we ranked the cards by their numerical value before separating them into suits, their shared status as "threes" would more closely link the two.

This is the same type of disclosure we see in our classification of life. Bottle-nosed dolphins and pompano dolphins appear very similar in several ways: they both swim in the sea, employ the same means of locomotion, eat the same type of food, and have comparable body shapes, tail fins, dorsal fins, and pectoral fins. Yet the bottle-nosed is a mammal and the pompano, a fish. And if one tracks their classification back to a common rank, it would be at the intermediate rank of subphylum, five ranks away, where the reptiles, amphibians, and birds also come together. So, in spite of their similarities, taxonomy shows us that the bottle-nosed and the pompano are only distantly related.

When we ranked the deck of cards, we did so according to a principle called *phenetic* classification, which relies on similarities. We first considered similar color, then suit, status as a court card or not, and whether it was evenly divisible by two or not. Phenetically, we could just as well have first divided the deck into court and noncourt cards instead of red and black. Yet in the majority of cases (those not involving most microorganisms), biologists are able to employ a much more precise and informative technique.

When ranking the higher organisms, scientists most often use *phylogenetic* analysis,[6] a technique that while relying on the similarity of shared structures and functions, also considers whether they are primitive or subsequently derived. Therefore, apparent close similarity does not automatically equal close relatedness. In practice, this prevents us from indiscriminately taking a trait—such as possession of a tail, for example—and classifying animals into tailed and nontailed groups. Such an exercise would put monkeys with alligators and humans with oysters. While this classification is logically valid, it tells us nothing about the real relationships among the four.

Along with shared traits, assessment may also involve patterns of evolutionary descent based on the elapsed time between common ancestors. Taken together, these factors and other considerations form the basis of all taxonomic groups, which are called taxa. These are the groupings that fill out the system of ranks extending from domain down to species, a reflection of evolutionary progress (or at least kinships), as best we understand it (domain, a new rank, is discussed later on).

Conceived in the eighteenth century by Carolus Linnaeus (1707–1778), the system originally classified life forms into three ranks: kingdom, genus, and species. Later on, Linnaeus would add the ranks of class and order, to which the German evolutionist Ernst Haeckel[7] inserted the ranks of phylum and family (botanists and bacteriologists often use "division" as an equivalent label to "phylum" in their classification of plants and bacteria).[8] Today, every form of life, from dinosaur to newly discovered microbe, is assigned a position within each of the eight ranks. For humans, lions, and alligators, these rankings are as follows:

	Human	**Lion**	**Alligator**
Rank	Taxon	Taxon	Taxon
Domain	Eucarya	Eucarya	Eucarya
Kingdom	Animalia	Animalia	Animalia
Phylum	Chordata	Chordata	Chordata
Class	Mammalia	Mammalia	Reptilia
Order	Primates	Carnivora	Crocodilia
Family	Hominidae	Felidae	Alligatoridae
Genus	*Homo*	*Panthera*	*Alligator*
Species	*sapiens*	*leo*	*mississippiensis*

As a hierarchy, this system of ranks is very simple. A genus is a group of closely related species whose members have more in common with each other than with members of species of other genera. A family is a group of closely related genera. An order is a group of closely related families, and so on, up to the rank of kingdom.

Notice that there is no rank called subphylum listed, which was mentioned as the first common rank of mammals, fish, amphibians, reptiles, and birds. This rank, along with others, is an optional fine-tuning of the hierarchical system that is used when a more precise grading of a lineage is required. If a particular trait within a taxon exhibits a significant split in evolutionary development, the rank may be subdivided into smaller units or may be subordinated under an optional higher rank.

In the ranking of *Homo sapiens*, both types of fine-tuning have been proposed: between the ranks of phylum (taxon, Chordata) and class (taxon, Mammalia), the ranks of subphylum and superclass can be established. The subphylum rank indicates that Chordata was subdivided, with the mammals, fish, amphibians, reptiles, and birds assigned to the subphylum taxon Vertebrata. (The other subphylum taxa are Urochordata, the sea squirts; Hemichordata, the acorn worms; and Cephalochordata, the lancelets.) The superclass rank indicates that Mammalia, along with other Vertebrata classes (Amphibia, Reptilia, etc.), were grouped into at least two higher taxa. There are two: Agnatha, for those without jaws (e.g., the lampreys); and Gnathostomata, for those with jaws. A more precise ranking of humans then is Animalia—Chordata—Vertebrata—Gnathostomata—Mammalia—Primates—Hominidae—*Homo*—*sapiens*. Other proposed human hierarchies have different or additional intermediate ranks.

In a few cases an intermediate rank is established simply as a convenience, rather than to express a substantiated relationship among organisms. The intermediate taxon Glires, a cohort rank, has been used by some to group the rodent order Rodentia with the rabbit order Lagomorpha. Lacking any concrete evidence of close kinship, Glires is based merely on the likelihood that in the distant past rodents and rabbits were closely related.

As with the eight main ranks, intermediate ranks follow an established order, with each scientific discipline determining its own groupings. For the animal and plant kingdoms, these groupings fill in the mandatory eight-rank hierarchy as follows.

Domain: Eucarya

Animals	Plants
Kingdom	Kingdom
Subkingdom	
Phylum	Division/Phylum
Subphylum	Subdivision
Infraphylum	Infradivision or Branch
Superclass	
Class	Class
Subclass	Subclass
Infraclass	
Supercohort	
Cohort	
Superorder	Superorder
Grandorder	
Mirorder	
Order	Order
Suborder	Suborder
Infraorder	
Superfamily	
Family	Family
Subfamily	Subfamily
Tribe	Tribe
	Subtribe
Genus	Genus
Subgenus	Subgenus
	Section
	Subsection
	Series
	Subseries
Species	Species
Subspecies; or Race, Breed	Subspecies
Form	Variety
Infrasubspecies; or Race	Subvariety
Breed, Form	Form
	Subform
	Cultivar

Any organism classified in an intermediate rank must be included in all higher, but not necessarily lower, intermediate ranks within the major rank. Therefore, if an animal's classification includes infraclass, it must also include class and subclass but not necessarily cohort. Those intermediate ranks designated as "super . . . " are not considered an extension of the preceding rank but of the next major rank. Superfamily, for instance, is a member of the "family group" of ranks, although it precedes the rank of family.

No organism has a lineage traceable through all of these intermediate ranks, but some plants do require quite a few. And whereas intermediate ranks do serve an important function, some have been of only minimal use, such as supercohort, grandorder, and mirorder. Certain intermediate ranks, such as series, are used only in botany, while others, like race, are used primarily in zoology. And not all ranks are officially recognized by taxonomic codes. In zoology, only the eight primary ranks and those prefixed by "super" or "sub" are officially recognized. In botany, the eight primary ranks and only those prefixed by "sub" are acknowledged. Moreover, no taxonomic code officially sanctions the ranks of breed or cultivar[9] ranks used for artificially created variants. Also, bacteriology uses the unofficial rank of strain to classify bacteria populations having a particular character, such as the biovars, the morphovars, the serovars, and the phagovars.

Biological classification is a rigorous undertaking, involving great care in the collection and interpretation of data. Unfortunately, the shortage, quality, and inconsistency of such data may not allow for anything more than theorizing. The result is that hierarchical trees are not immutable facts but assumptions, and the placement of an organism or its extended lineage often rests on supposition. This is why one will find instances throughout this book where authorities are in disagreement. Entomologists, for example, often assign the cockroaches to the suborder Blattaria and the mantids to the suborder Mantodea, both of which fall under the order Dictyoptera; however, other authorities eliminate Dictyoptera as rank and treat Blatteria and Mantodea as independent orders, with no change in name.[10]

But an even greater challenge confronts the bacteriologist. Until recently, this relatively new field was unable to identify any phylogenic relationships among bacteria; consequently, species were ranked only by phenetic criteria, which means that many will now be in need of substantial reclassification. Fortunately, other fields of study have eluded such a broad reorganization, yet because their taxonomy is far from complete, they are not without their own classification problems and challenges.

The root of such disputes is not that one scientist is less informed than another—although this certainly could be the case—but that the whole task of classification is hindered by inadequate and conflicting data. To deal with these obstacles, scientists constantly wrestle with an assortment of questions: What traits are significant? What weight should they carry? Are these primitive traits or are they later modifications? Are they shared by all members of a group? Such questions typically result in answers that not only shape the process of classification but also determine the foundation of each taxon. This is particularly true of those ranks extending from phylum/division to genus. Unlike the rank of species, they are not governed so much by definition as they are molded by the traits and evolution of their subjects, which often creates the need for intermediate ranks.

Because the foundation of each taxon is determined by the characteristics of its members, the magnitude and the extent of the defining traits among various taxa within a rank can vary widely. Within the bat order, Chiroptera, the family Phyllostomidae is distinguished from the family Vespertillionidae only by an additional bone in the third finger and often the presence of a leaflike projection on the nose; whereas in the Artiodactyla, even-toed ungulates, a considerable number of differences distinguish the pig family Suidae from the giraffe family Giraffidae. Furthermore, within a taxon not all relevant identifying characteristics may be applicable to all members. In a particular order, the traits involved in separating one family from another may not be the same as those that distinguish it from a third family.

This illustrates an important point: individual taxa of a rank seldom share a common foundation. For example, within the rank of class the principles that define the clam taxon, Bivalvia, are nothing like those that define the millipede taxon, Diplopoda. But, then, what defines a rank such as class? What is the common denominator that makes a class a class or a family a family? Well, there really isn't any. Each rank is only a position or a level, not a biological construct. Scientists are constrained by the characteristics and the evolution of their subjects, while at the same time are bound by taxonomic rules. In practice, no matter how strange or unique a new organism may be, characteristics are almost invariably established (usually phylogenetically, as much as possible) to suit each of the mandatory eight ranks. One current exception is the amoeba-like animal trichoplax, for which no class or order has been established. Taxa of higher ranks are generally defined by the more general or gross features of an organism. Lower-ranked taxa like genus and species depend on more particular features. When a ranking is completed, there will be eight specific levels, each defined by increasingly distinctive features as the species rank is approached.

Yet sometimes, as with a unique organism, formulating the specifics for each rank is not only irrelevant, but the resulting taxa are essentially meaningless. As an example, the ginkgo tree is commonly listed as belonging to the class Gymnosperma, order Ginkgoales, family Ginkoaceae, genus *Ginkgo,* and species *biloba.* It is the only species in the entire order, which means there is essentially no useful difference between the characteristics that define its order taxon and those defining its species taxon (the same is true of the aardvark, the singular representative of the mammalian order Tubulidentata). In such cases the formation of a taxon may be no more than a formality as required by the rules of taxonomy. And, as with the ginkgo order, in such instances intermediate taxa often go unmentioned in literature.

Of the eight ranks, kingdom and species are decidedly unique and deserve elaboration. When Linnaeus devised his system incorporating genus and species, three great kingdoms were recognized: animal, plant, and mineral. Biologists, ignoring the mineral kingdom for obvious reasons, decided that if an organism could move about and digest food, it belonged in the animal kingdom, and if it stayed put and did not digest food, it went into the plant kingdom. However, with the advent of more powerful microscopes and exacting biological techniques, it was soon discovered that some organisms, mainly microorganisms such as bacteria and amoebas, were neither plants nor animals. This led scientists to seriously question the two-kingdom classification as it was then defined. More important, it became obvious that life required more than two kingdoms to properly classify its diversity. But the ensuing task of deciding what kingdom an organism belongs to was not an easy one. Consequently, classification systems with between 4 and 20 kingdoms were suggested, each with varying advantages and disadvantages.

This dilemma was tentatively resolved in 1959 when R. H. Whittaker proposed a concept employing five kingdoms, a system that soon became widely accepted. Like others, Whittaker based his kingdoms more on cellular structure and nutritional mode than on the evolutionary principles that underpin the other ranks. His five kingdoms, with examples, are:

Monera	Bacteria and cyanobacteria
Protoctista	Amoebas, algae, ciliates
Fungi	Mushrooms, mildews, yeasts
Plantae	Trees, flowers, dandelions
Animalia	Clams, bees, humans

As suitable as this system was, some organisms resisted rigid classification. Slime molds, for instance, possessing both animal and plant characteristics, were, until quite recently, claimed in both Protoctista and Fungi. In a more recent scheme (1978), the five kingdoms were grouped into two superkingdoms: Eukaryotae, organisms with nuclear membranes, for example, plants and animals; and Prokaryotae, organisms without nuclear membranes, the bacteria. (A nuclear membrane is a covering that surrounds the nucleus of a cell.) In effect, this reordering gave us two major branches of life.

However, this distinction lasted only until 1996, when the existence of a third branch of life was said to be proven—the branch from which all other forms of life are thought to have arisen. Named Archaea, its microbial members were once known as archaebacteria and were classified as a distinct group within Monera.[11] The Archaea, a very peculiar group of organisms, are distinguished not only by their genetic makeup but by where they live and how they function. Often found living on sewage and in sediments and swamps, they exist without sunlight, feeding on carbon dioxide, nitrogen, and hydrogen and producing methane gas as a waste by-product. They also live in places hostile to all other forms of life: 8,600 feet under the ocean at pressures 200 times greater than on Earth's surface, in 185° F water (close to boiling), and in extreme cold. But more important, their distinction rests in the fact that two-thirds of their genes do not resemble those found in any other form of life.

The upshot of all this reassessment and reworking is that taxonomy has now been expanded to eight ranks. At the top is the rank of domain, which consists of three taxons: Bacteria, Archaea, and Eucarya (although the rank of domain has been in use for several years, it has still not been universally accepted). In one revision the rank of superkingdom has been dropped, Eukaryotae was renamed Eucarya, and Prokaryotae has been eliminated. Also, the kingdom Protoctista is no longer a formal designation, although it is sometimes used to indicate all members of Eucarya, other than those belonging to the Fungi, Plantae, and Animalia kingdoms. Those organisms that once made up Protoctista have now been divided among as many as eighteen new kingdoms (some have also been reassigned to the Fungi, Plantae, and Animalia kingdoms), each on equal footing with Fungi, Plantae and Animalia. All of this reordering is quite new and much of it is tentative, and, of course, not all scientists are satisfied with it, so there are certain to be many changes before taxonomy really settles down.

The following is an example of one newly proposed reordering. Because the Bacteria domain is quite unsettled, the validity of its kingdoms, listed in the table, is very tenuous. The Archaea domain is also shaky, with additional kingdoms likely to surface in the future. So, too, with Eucarya; with continued molecular studies, it will certainly undergo further revisions.

Domain: Eucarya

Kingdoms			
Animalia	Plantae	Fungi	Diatoms
Phaeophyta	Chrysophyta	Xanthophyta	Oomycota
Labyrinthulids	Apicomplexa	Dinoflagellata	Ciliates
Rhodophyta	Acrasiomycota	Entamoeba	Naegleria
Euglenozoa	Myxomycota	Parabasalids	Microsporida
Diplomonads			

Domain: Archaea

Kingdoms		
Euryarchaeota	Crenarchaeota	Koryarchaeota

Domain: Bacteria

Kingdoms			
Proteobacteria	Planctomyces	Chlamydia	Spirochaetes
Bacteroids	Firmicutes	Thermotogales	Hydrogenbacter
Cyanobacteria and Chloroplasts		Green sulphur bacteria	

Thirty-four kingdoms![12] This is an incredibly explosive increase in just a single reordering, and with more likely on the way—not a comforting thought for the interested layperson or the beginning biology student. And aside from the future revisions that taxonomy is bound to undergo, most of the kingdoms, besides being saddled with tongue-twisting names, will seldom be remembered or understood by anyone other than biologists. Perhaps, if the layperson is lucky, most of these kingdoms will eventually be reduced to subkingdom status or combined with one another. But this is a very big "perhaps."

At the species level, the critical differences separating one species from another often range from the obvious to the obscure. Among the lizards of Florida, it is easy to spot the external features that distinguish the bright green–colored green anole *(Anolis carolinensis)* from the somber-hued brown anole *(A. sagrei),* but for the gray wolf *(Canis lupus)* and the red wolf *(C. rufus)* there is little apparent difference except a slight variation in size and coloring (some authorities have suggested that the red wolf be regarded as a subspecies of *C. lupus).* And when one looks at dogs (domestic canines, *C. familiaris),*[13] such traits are totally ignored, allowing a range of shapes, sizes, and colorings that is immense. This illustrates why outward appearances can be deceiving. The Alaskan malamute, for instance, appears to have more in common with the gray wolf than with the chihuahua, but when their DNA profiles are considered, and from an exacting anatomical standpoint (mainly, skull features), this is decidedly not the case. What does separate one species from another usually rests with reproductive factors—the crux of the species rank and much of the controversy that surrounds its definition.

Unfortunately, the biological sciences have yet to come up with a universal definition of "species." The various species concepts now in use have arisen out of the need to address classification issues that are often particular to only one branch or sphere of biology and therefore are seldom widely applicable. These concepts include biological, biosystematic, evolutionary, genetic, morphological, paleontological, and phylogenetic definitions.

The most widely used of these definitions, based on the biological concept of species, usually takes a form akin to that formulated by biologist Ernst Mayr. Mayr contends that a species is a "reproductively isolated aggregate of populations which can interbreed with one another because they share the same isolating mechanisms."[14] The heart of this definition lies in the criterion of reproductive isolation.

Reproductive isolation embraces any element of nature that prevents the reproduction of fully viable offspring. Some of these elements keep organisms apart before they can mate. These include ecological barriers, which physically keep populations from meeting; temporal barriers, wherein species breed at different times of the day or the year; behavioral barriers, where organisms consciously reject, for whatever reason, other organisms; and mechanical barriers, such as incompatible physical or physiological differences in the two organisms. Other prohibitive elements occur after mating has taken place. These include cases where the sperm and the egg do not reach each other or fail to fuse; where a hybrid dies before it sexually matures; and where the hybrid is sterile. The essence of this definition of species is the ability of two organisms to successfully reproduce by mating.

However, not all organisms reproduce by mating. Some asexual organisms reproduce by fission, where an organism simply splits into two equal organisms (euglenids); binary fission, in which a parent organism splits into two offspring (certain green algae); and parthenogenesis, in which there are no males—the egg always develops into a female (bdelloid rotifers).

Although most flowering plants reproduce sexually, there are those that self-pollinate, so the use of the biological definition of species to sort out these particular plants is questionable. Also, the biological definition cannot be applied to fossils. Obviously, in such cases other criteria must be used to determine species. Typically, these will involve external morphology (physical characteristics), chemical and physiological properties, or genetic makeup. A species that reproduces by mating is also usually considered to be a naturally occurring group of interbreeding, or potentially interbreeding, organisms that shares a unique gene pool and produces fertile offspring. Within this somewhat qualified statement are several key ideas.

"Naturally occurring" excludes offspring resulting from species deliberately brought together. In the wild, Bactrian camels *(Camelus bactrianus)* and Arabian camels *(C. dromedarius)* do not mate because their dispositions, habits, and territories differ; consequently, their gene pools go unmixed. Yet in Turkey, where they are deliberately interbred, they produce fertile offspring (some males are born sterile).

Not many crossbreedings between closely related species are this successful. If offspring do result, they are most often either deformed or infertile. (One glaring exception to this stipulation is *Canis,* the genus of wolves, coyotes, and dogs—see "Wolf," chapter 7). Mules are such sterile animals, the product of the mating of a mare and an ass—there are no mother or father mules. Yet not everyone feels that the fertility/infertility criterion of species definition is a practical one. Keith Rushford, in his book *Conifers,* observes, "The traditional criterion of a species being interfertile between members of the species and, under natural circumstances, at least partially intersterile with members of another species is subject to too many qualifications to be useful in practice."[15]

But "interbreeding" does keep intact those features that distinguish one species from another. Although some cross-species mating may take place, usually it is not enough to break down the species' identity. The major factors in preserving a species' identity are the isolating mechanisms mentioned in the biological definition of species.

"Potentially" is used to include within a species those populations that while not normally interbreeding, could successfully do so. The

Inuits of the North and the Pygmies of Africa do not mate with each other, yet they are not considered separate species because, among other reasons, unlike the two camel species, there are no instinctual barriers inhibiting mating. Humans readily disregard "race" and often find those of different ethnicities quite desirable. Bactrian and Arabian camels, given the choice of mating within their species or with each other, prefer to mate with their own—they are not considered to be potential interbreeders. Within a species, those unique populations of potentially successful interbreeders are often designated as subspecies or, as was the case with humans, races.

A subspecies is almost always a geographically isolated group, which, given enough time (usually, thousands of years), may evolve into a bona fide new species. Crucial to such changes are shifts in the environment and the accompanying genetic mutations that offer the necessary variations needed for adaptation.

Although "subspecies" sits at the bottom of the hierarchical ladder, its importance should not be underestimated. "Kingdom" may have greater status and "species" more recognition, but it is "subspecies" that is often the laboratory of incipient species, the rank of innovation, and the genesis of diversity.

Various mnemonic sentences have been devised to help you remember the order of the eight ranks. Here are three that may come in handy.

Domain	Kingdom	Phylum	Class	Order	Family	Genus	Species
Daffy	King	Philip	Came	Over	For	Ginger	Snaps
Daddy	Keeps	Putting	Cashews	Out	For	Gray	Squirrels
Danish	Kings	Play	Chess	On	Furry	Green	Squares

2
PLANTS

Bulb ● Corm ● Tuber ● Root

Bulbs, corms, and tubers, unlike roots, are modified portions of underground stems. As with the portion that grows above ground, they have nodes and axillary buds. A node is a point on a stem where buds, leaves, and branches form; an axillary bud is a bud formed in the angle between the stalk of a leaf and the stem from which it grows. Roots have no such features. However, one characteristic that all four do share is the ability to store food. For most, it is their main function, one that provides new spring growth with nourishment after a season of dormancy. Confusion among the four usually occurs between bulbs and corms and between roots and tubers.

The bulb is distinct from the other stem forms because it is a highly modified bud comprised almost entirely of fleshy leaves. If these leaves tightly wrap themselves around the embryonic plant located at their center, they are called a tunic. Onions and tulips have bulbs of this type, with thin, papery leaves. Daffodils also develop a tunic, but their leaves are a bit thicker. So, too, are those of Easter lily bulbs, which develop as a looser sheath of leaves that resembles pointed scales. But regardless of the type of leaf, bulbs all contain sugar, starch, and proteins, essential nutrients for growth. On the bottom of the bulb is a thickened platform called the basal plate, from which its leaves emerge. Roots also issue from around the plate's edges, and new bulbs, termed bulblets, arise laterally from its top. If the plant is confronted with a poor growing season, the reserve food in the bulb will not be used for growth but for survival, and the plant may fail to flower.

The corm, despite its superficial resemblance to the bulb, lacks fleshy leaves; however, it does have a few papery outer leaves that act as a protective covering. Also, corms are usually squatter than bulbs, although some, such as those of the crocus, are round. And, like the bulb, the corm has a basal plate from which roots issue.

In the growth cycle of the corm, buds readily sprout from nodes dotting its upper surface. Competition among these buds for nutrients eventually results in the formation of one or two dominant corms. The other buds, producing smaller corms called cormels, do not blossom the same year but will sprout two or three years later. Because the development of buds into new corms and cormels draws heavily upon stored food, such reserves are ultimately exhausted, and the parent corm eventually shrivels up and dies.

Tubers, such as the common potato, are easily distinguished from bulbs and corms by their lack of a basal plate and enveloping leaves. Lacking leaves to prevent moisture loss, tubers develop a tough skin, and their roots, which do not require a specialized anchoring structure, are apt to sprout from anywhere. Tubers emerge from the ends of horizontal underground stems called rhizomes or from above-ground stems, in which case they are called aerial tubers. As the largest of all stem storage organs, some tubers can swell to over 125 pounds. Typically, they take on a rounded "potato" shape, although a few, like the begonia, may be flattened or, like the anemone, knobby. Probably the best-known feature of the tuber is its nodes, or "eyes." The bulb and the corm have nodes as well, but only on the tuber are these conspicuous. Actually, an eye, nestled in the remnant of a scale-like leaf, is an axillary bud from which a new shoot will form.

Roots are seldom confused with bulbs, corms, or tubers unless they occur as a swollen storage organ. The most notable misidentified root is the sweet potato. Because it resembles the potato tuber in shape and texture, and its name in part is identical, the mistake is understandable. But when large seed companies pass along such misinformation in their catalogues, it is unpardonable. One company in particular repeatedly calls its sweet potatoes "tubers."[1] This is puzzling because even a passing examination of the vegetable reveals it is not: it has no eyes. No roots do. The only growth bud on this tuberlike root is at the root's crown, the base of the old stem. Only from this point can a new shoot emerge. Sometimes called tuberous roots, these enlarged sections of a root do not take in water and nutrients themselves but rely on much smaller attached fibrous roots. Along with the sweet potato, the dahlia root is also often mistaken for a tuber.

Corn • Maize

Most schoolchildren who have learned of the Indians and their agriculture know that one of the more popular Native American words for corn was "maize," which means "giver of life." This is not its original spelling; "maize" derives from the Spanish *mahis* or *mahiz*. It was a word the Spanish first heard when they encountered the Taino tribe of Arawakan Indians on Hispaniola in the West Indies. The Spanish borrowed the term because they lacked any other; maize was a new grain never before seen by Europeans until their arrival in North America. A staple food of the Indians, it is thought to have been domesticated over 7,000 years ago from a large wild grass called teosinte.[2] Although teosinte still grows wild as an annual, maize has been so highly bred away from its origins that it is unable to reproduce on its own in the wild. Because of its status as a basic food source, it was continually nurtured and improved, and it is now widely grown and known by many names.

Today in South America, the old Incan word *choclo* is still used for corn on the cob. Elsewhere in South and Central America, "mealies," a term based on the Portuguese *milho* and brought over from Africa by slaves, remains popular. Ironically, the word "maize" followed the introduction of corn to Africa, where it became the prevailing term. After the arrival of maize in India, it was confused with the sorghum plant in many areas and took on the name "sorghum."

During this era of exploration, explorers from other countries—principally England, France, and Portugal—also encountered the new grain, later given the botanical name *Zea mays* (*Zea,* from the Greek *zeia,* for "grain"). The English, unfamiliar with the new Spanish name and reluctant to use a native word for the new grain, called it "corn." Other settlers of North America, who eventually came under British rule, soon adopted the term, along with the qualifier "Indian." Thus, since 1608, the grain's new English name was "Indian corn,"[3] later shortened to simply "corn." But the question arises: why bother adding the term "Indian" in the first place?

North Americans readily use the word "corn" by itself because it refers to only one grain, *Zea mays.* In Europe, however, the term has had a quite different meaning, one rarely adopted in the New World. In England, Scotland, Ireland, Germany, France, and Italy, "corn," in its various spellings, often refers to either the principal cereal grain of a country or simply to grain in general (the equivalent term in Spanish is *panizo*).

"Corn" comes from the Old Saxon *korn,* which in turn was taken from the Old Teutonic *kurnom,* a word akin to the Latin term for grain, *granum*. Originally, the word was applied to any hard seed or grain, which is the sense in which it was used for the New World's maize. "Indian corn," then, meant "Indian grain." Likewise, "John Barleycorn," the whimsical expression given to malt liquor, refers to barley grain—not to corn, as the word is now used in the United States.

Today, Europeans tend to define "corn" a bit more narrowly, often using it to denote specific crops. In England, although "corn" may be used for any cereal crop, it is frequently limited to just wheat; in Scotland and Ireland, to oats; in parts of Germany, to rye; in Italy, to either wheat or oats; and in France, to any cereal or grain. But many older Europeans continue to use "corn" with its broader meaning, applying it to rice, barley, millet, sorghum, maize, and in the broadest sense to beans and peas.

It is this European use of "corn" for such various grains, cereals, and crops that is most apt to confuse North Americans. However, if there is a possibility of misunderstanding when speaking of *Zea mays,* most Europeans will use the term "Indian corn" or "maize." Therefore, in Great Britain cornmeal is called either maize flour or Indian corn; corn on the cob, either sweet corn cob or maize cob; and crushed corn for cattle, maize flakes. The Portuguese use the word *milho* for maize, which comes from their word *milhete* for the grain millet. And in France maize is still sometimes referred to as "Turkey wheat," a name connected to its former Latin name *Turcicum frumentum*—both terms having been based on the mistaken notion that the plant was brought out of Asia by the Turks.

Evergreen ● Deciduous ● Broadleaf ● Coniferous

Trees are usually perceived as belonging to one of two broad categories: the evergreen conifers and the deciduous broadleaves. "Evergreen" denotes year-round green foliage, "coniferous" alludes to the cone shape of a tree's seed-bearing structures, "deciduous" indicates that leaves are shed seasonally, and "broadleaf" refers to wide, flattened leaves, in contrast to needles. Although the terms "evergreen conifer" and "deciduous broadleaf" are an easy way to categorize trees, other combinations of these four terms are equally valid. And because these terms are sometimes used rather loosely, they may not accurately represent a tree's true character. So, whereas most trees are either evergreen conifers or deciduous broadleaves, exceptions do exist: four in particular are noted in the following.

Botanically, seed-bearing plants take two forms: the angiosperms, which sprout flowers and produce seeds enclosed in an ovary; and the gymnosperms, which have no flowers and whose seeds are naked. Gymnosperms are best represented by the conifers of Coniferophyta, the largest of four divisions, and whose seeds are borne on the wings of their female cones. (The other divisions are Ginkgophyta, the ginkgo; Cycadophyta, the cycads; and Gnetophyta, such as the welwitschia plant.) However, (first exception) not all those trees we call conifers bear true cones.

The yew, long called a conifer because of its outward appearance and sometimes mistakenly said to bear cones, does not produce cones but arils. These are bright red fleshy cups that hold a single greenish seed. And whereas this "fruit" may look like the product of a flowering angiosperm, it still lacks the necessary ovary wall. The same is true of the seed of the podocarp, a conifer of the Southern Hemisphere. Unlike the aril, these blue-gray seeds have an appendage and are not open on one end.

Except for the ginkgo tree, which is a gymnosperm but not a conifer, all other trees bear flowers and have enclosed seeds. These trees are termed "broadleaves," a distinction meant to imply exclusion from the gymnosperms and inclusion with the angiosperms. However, (second exception) "broadleaf" also aptly describes the flat leaves of the ginkgo.

Palms, those stately trees with broad fan- or feather-shaped leaves, are generally not considered broadleaf trees either. Unlike the other angiosperms, which are designated as dicots because of the two leaves each of their seeds bear, palms are monocots (like the grasses), whose seeds each bear only a single leaf.

With the exception of the palms and their close relatives, all other flowering trees are "broadleaves," a category that includes the oak, the apple, the poplar, the basswood, the teak, and the eucalyptus. These six particular trees are also deciduous: their leaves fall off with a change of season. While we normally associate the approach of winter and lower temperatures with the onset of leaf shedding, in warmer tropical climates deciduous trees also flourish. Trees in these regions respond not to temperature change and shortened daylight but to lack of rainfall. A good example is the teak tree of the monsoon forests of Southeast Asia, which drops its leaves every year when the five-month-long dry season sets in.

Are all deciduous trees broadleaves, then? No, (third exception) in the north during the later months of winter, the tamarack or North American larch, a conifer, will stand with branches bare, its brown needles blanketing the ground below. Two other conifers, the bald cypress of the south and the metasequoia of China, are also deciduous. So, whereas the term "deciduous" is usually taken to be synonymous with "broadleaf," there are the exceptions: not all conifers are "evergreen."

"Evergreen" is a term we usually reserve for conifers and sometimes, inappropriately, for the three deciduous species just mentioned. This does not mean, however, these nondeciduous trees never lose their needles. Needles (a form of leaf) eventually tire and die, at which time the tree must shed them to make room for new ones. Unlike deciduous shedding, this process does not necessarily occur seasonally but rather at various times throughout the year, which varies with each species. The average conifer sheds only 10 to 20 percent of its needles each year, so the average life of a pine needle, for example, is 2 to 4 years, whereas needles of the bristlecone pine are retained for up to 45 years. Furthermore, (fourth exception) it is a mistake to regard conifers as the only evergreens.

Broadleaf trees, such as the acacia, olives, certain oaks, and some magnolias, also hold on to their leaves for more than a year. Depending on the species, a holly can be deciduous or evergreen. And in the tropical rain forests, where the temperature and the rainfall are consistent, broadleaf trees with permanent foliage abound.

From the foregoing, it is apparent that along with the typical broadleaf deciduous trees and the evergreen conifers, we also have:

Broadleaf evergreens (acacias)
Nonflowering broadleaves (gingkoes)
Non–cone bearing conifers (yews)
Deciduous conifers (tamaracks)

Fruit ● Vegetable

Is the tomato a fruit or a vegetable? Strangely enough, in 1893 the United States Supreme Court ruled on this very question. Tomatoes, it concluded, are vegetables. Botanists, while conceding that the designation is fine for commerce, will point out that botanically, a tomato is a true fruit in every sense of the word (more specifically, it is a berry). The court—involved because importers disputed the tariff placed on tomatoes (then deemed vegetables and therefore taxable, rather than fruits, which were not)—based its ruling primarily upon the observation that vegetables are herbaceous plants or plant parts we serve at the table as a salad or the main course; whereas fruits, being sweeter, are reserved for dessert or are often eaten alone. Tomatoes are not particularly sweet nor are they served as a dessert. This view that the court held was certainly not a scientific one, which is why confusion over the two terms was almost inevitable.

In botany, the word "vegetable" has very little meaning in describing plants, other than to suggest their horticultural nature or culinary use. A vegetable may be loosely defined as any plant or plant part, other than a sweet fruit, suitable for eating. But "fruit," a true scientific term, describes a very specific part of a plant—edible or not. Botanically, a fruit is the ripened, *sometimes fleshy,* ovary of a flowering plant that surrounds its seeds. It is quite obvious what this means when we look into an apple, a peach, or a grape and see its seeds nestled in pulp. But what about strawberries and pineapples? Looked at closely, it may be difficult to see how their construction fits the definition of a fruit.

In the strawberry, many separate ovaries sprout from a single receptacle of *one* flower. The receptacle is the large red fleshy body we prize, and what we often call the seeds that nestle in its dimpled surface are actually individual fruits. This type of fruit is termed an "aggregate." The pineapple, a second type of fruit, is a "multiple." Multiples form from the individual ovaries of several flowers. On the pineapple, each raised button on its surface is formed by a separate flower and is called a fruitlet. A third type of fruit is the simple fruit. Simple fruits like apples and peaches have only one ovary, as in the brief description previously given.

The following list shows where various fleshy fruits fit into these three categories.

Simple (one flower, one ovary)		Aggregate (one flower, many ovaries)	Multiple (many flowers, one ovary each)
apple	banana	strawberry	pineapple
plum	cherry	blackberry	mulberry
date	quince	raspberry	osage orange
grape	lemon	dewberry	fig

Unquestionably, these are all fruits; however, consider the following fruits that we normally speak of as vegetables:

corn	okra	cucumber
sunflower	walnut	pepper
buckwheat	rice	acorn
sugar pea	coconut	eggplant
tomato	chestnut	almond
string bean	pumpkin	olive

It may be surprising that buckwheat, the acorn, and the almond are all fruits, yet they all fall under the definition of a simple fruit—recall, not all fruits are fleshy. Even the twin-winged "seeds" of the maple tree that spin as they fall to the ground and the wispy parachutes of the dandelion are really fruits—dry fruits containing a seed. A few others in the list deserve some clarification. The sugar pea and the string bean, as picked from the plant, are fruits; however, the individual peas and beans are not; they are its seeds. Likewise, the hard, thin shell of the coconut filled with white "meat" and liquid is a seed, but when surrounded by its fibrous husk of copra and protective rind, it makes up the coconut fruit. Incidentally, the coconut, like many other so-called nuts (e.g., peanut, pine nut, and pistachio), is not a true nut.

Because simple fruits come in such a variety of forms, botanists have found it convenient to classify them in more specific groupings. The following descriptions are not necessarily botanical definitions but are the more easily observed differences common to each group (this listing is not complete; it presents only the more common forms of simple fruits).

Fleshy simple fruits

Berry	Fleshy fruits, usually with numerous scattered seeds (currant, gooseberry, grape, tomato)
Hesperidium (citrus)	Modified berries with juice sacs and a thick rind that is separable (lemon, lime, orange, grapefruit)
Pepo	Modified berries with an inseparable hard, thick rind (cucumber, pumpkin, squash, watermelon)
Drupe	Fleshy or fibrous fruits with a thin skin and a seed enclosed in a stoney shell (almond, cherry, coconut, olive, plum, peach)
Pome	Fleshy fruits with a thin skin and seeds enclosed in a soft capsule (apple, pear, quince)

Dry simple fruits that rupture to release their seeds

Capsule	Dry fruits with multiple seed compartments (horse chestnut, lily, okra, poppy)
Follicle	Dry fruits with a single seed compartment that splits open along one seam (columbine, delphinium, magnolia, milkweed)

Legume	Dry fruits with a single seed compartment that splits open along two seams (soybean, pea, peanut, alfalfa)
Silique	Dry fruits like legumes, but with a two-segmented pod (cabbage, radish, mustard)

Dry simple fruits that do not rupture

Achene	Dry fruits with a single seed that attaches to its ovary wall at a single point (dandelion, sunflower)
Samara	An achene with winglike appendages (ash, elm, maple)
Schizocarp	Dry fruits with two fused seed compartments that separate at maturity into one-seeded halves (carrot, celery, parsley, parsnip)
Caryopsis	Dry fruits with a single seed that attaches completely to its ovary wall (the grains and the grasses: barley, corn, oat, rice, wheat)
Nuts	Dry fruits with a single seed enclosed in a hard, woody husk (acorn, chestnut, hazelnut, walnut)

Our convention of categorizing many of these fruits as vegetables is one of custom based on use, rather than on any rationale that makes a statement about the nature of the plant. That we may not have been aware these were fruits should not be surprising. How many of us recognize broccoli as a flower cluster? Or know that the brussels sprout is a bud? Or, whereas a russet potato is a tuber, a sweet potato is a root?

Vegetables, too, take on many forms. Whereas a fruit can be the product only of a flower, a vegetable may come from any part of a plant: root, tuber, bulb, stem, stalk, leaf, flower, bud, seed, rhizome, or fruit. The following, classified by these edible plant parts, are some of our more common vegetables.

Root	Beet, carrot, chicory, radish, rutabaga, salsify, turnip, horseradish, celeriac, sweet potato
Tuber	Potato, yam, taro, cassava, Jerusalem artichoke
Bulb	Onion, shallot, leek, garlic
Stem	Asparagus, kohlrabi
Stalk	Celery, rhubarb

Leaf	Lettuce, cabbage, parsley, spinach, chive
Flower	Broccoli, cauliflower, globe artichoke
Bud	Brussels sprout
Seed	Lima bean, peanut, soybean
Rhizome	Turmeric, ginger
Fruit	Pepper, squash, okra

Grass ● Reed ● Sedge ● Rush

Although rushes, sedges, and other similar-appearing plants are often spoken of as grasses, none belong to the grass family Poaceae (Gramineae), which contains both grasses and reeds. Sedges and rushes, distinctly different plants, belong, respectively, to the Cyperaceae family and the Juncaceae family. Although not grasses, they often fall under the umbrella term "grass," if not by mistake, then by our attempt to distinguish these thin-bladed plants from the broad-leaved forbs.

The family of true grasses, Poaceae, is one of the most important sources of food in the world for humans and livestock. It includes the cereal grasses: corn (maize), oats, barley, rice, wheat, rye, sugar cane, millet, and sorghum; and the forage grasses used by animals: timothy, buffalograss, ryegrass, and fescue. Of all the grasses, only bamboo develops a woody structure and attains tree heights, although a few, such as corn and sugar cane, can reach heights of 10 to 15 feet.

Grasses are most easily distinguished by their stems and leaves. The grass stem is cylindrical, jointed, and, with the exceptions of corn, sugar cane, and sorghum, hollow between its solid joints. Grass leaves occur in two ranks; that is, they emerge from the stem only along two rows running up the sides. These leaves also consist of two distinct segments: the lower part, called the sheath, wraps around the stem; the more familiar upper portion is the blade. One feature unique to almost all grasses is the ligule, a short growth at the juncture of the leaf and the stem, which looks like a truncated leaf. Neither sedges nor rushes have ligules. Another telling characteristic of grass is its flowers.

As a rule, flowers typically bear four modified leaf forms: stamens, which produce pollen; carpels, which form ovaries; sepals, protective outer coverings; and petals, their showy outgrowths. Although grass flowers have stamens and carpels, they lack sepals and petals. Instead, the flowers of grass are very compacted and are protected by several green scales. The flowers themselves are quite small and occur on spikelets; they produce dry seedlike fruits called grain or, in the case of some bamboo, berries.

Reeds are aquatic grasses that occur in several genera, but most notably in *Phragmites,* the "reed genus." Growing in swamps, along the wet shores of lakes and streams, and in brackish backwaters, they range from northern Canada to southern South America, as well as Africa, Australia, and Eurasia.

The four species of *Phragmites* range in height from 5 to 15 feet and arise from horizontal stems that may grow above or below ground. In some species these serpentine runners can reach lengths of 30 feet. What sets reeds apart from the other grasses are their smooth, rigid stems and, most significantly, their flowers, which occur in plume-like tufts with long, soft, silky white "hairs." Whereas this genus contains the most typical and common reeds, other genera contain similar reedy plants, such as the giant reed, the sea reed, and the blue joint reed. Even other families contain members called reeds, such as the bur reed and the reed mace; however, these are not true reeds.

Sedges, like reeds, grow in marshes, along shores, and in wet meadows throughout the world, particularly in temperate regions. Over 1,000 species form the immense genus *Carex* within Cyperaceae, which is better known as the sedge family. Similar to the grasses, sedges also lack sepals and petals, and their flowers are borne on spikelets atop the stem. Unlike the grasses, the fruit of the sedge is not a grain or a berry but an achene, whose seed is attached to the ovary wall at only one point (in grain, the seed is attached all the way around). Also, unlike grass flowers that contain both male and female reproductive structures, sedge flowers are unisexual. But what really sets the sedge apart from all the other grasslike plants is the unique triangular stem typical of most species, which is also solid and, unlike grass, lacks joints. In addition, the sedge leaves are three ranked on the stem, not two, as on grass.

Rushes are also aquatic plants, typically found in shaded marshy areas and lake margins of the northern hemisphere's more temperate regions. The rush family Juncaceae has two principal genera: *Luzula,* the wood rushes; and *Juncus,* the common rushes. *Juncus,* by far the larger genus with about 225 species, contains the rushes used throughout the world for weaving mats, chair bottoms, and the like. Unlike grasses and sedges, the flowers of the rush embody sepals and petals. The flowers are small, sometimes inconspicuous, and form terminal brown or greenish clusters called cymes. The seeds of the rush develop in small pods, not individually as in the other plants. Like the grasses, the stems of the rushes are round; however, usually they are not completely hollow but rather are pithy and sometimes partitioned horizontally.

As with the term "reed," "rush" has also been misapplied to a number of plants. Sweet rush, spike rush, beak rush, nut rush, and the well-known bulrush are not true rushes at all but plants of the sedge family. Likewise, "rushes" such as the scouring rush, the flowering rush, and the sweet flag rush are all plants belonging to families other than Juncaceae.

Hardwood ● Softwood

The lumber industry and the woodworking trade readily recognize the differences between softwood and hardwood lumber. More than just a degree of hardness, each class of wood has its own structural, finishing, and processing properties.

Softwood and hardwood trees are distinguished by their inclusion in one of two different botanical groups: gymnosperms (typically, needled, cone-bearing plants) and angiosperms (broad-leaved, flowering plants). We commonly equate the conifers with the softwoods and the flowering trees with the hardwoods, regardless of their wood's true hardness.

Unfortunately, because hardness is used as an expression to designate the two classes, it is often assumed that all needled trees have softer wood than all broad-leaved trees. While this is typically true, there are exceptions. The hardwoods red alder, balsa, yellow poplar, aspen, and cottonwood all have softer wood than the eastern red cedar, longleaf pine, loblolly pine, coast Douglas fir, and western larch—all conifers. The convention of using "softwood" and "hardwood" to identify a tree is therefore not necessarily a true indication of its hardness but simply of whether it is a conifer or a flowering tree. Excluded from the broad-leaved hardwoods are the palms and the yucca, which have a very different type of stalk tissue and belong to another form of plant entirely, the monocots.

Conifers predate broad-leaved trees—in fact, all flowering plants—by millions of years. Ancient conifers existed over 200 million years ago during the Permian period, whereas flowering plants did not evolve until 130 million years later, during the Early Cretaceous period, and flowering trees in particular, later yet. This explains, in part, why softwood is a much simpler and less-efficient wood, and why it is not as durable as hardwood. In essence, it is more primitive, but this certainly does not mean the softwoods are any less desirable.

Softwood is very homogeneous; 80 to 95 percent of its volume consists of only one type of cell, the tracheid. These long, narrow, and hollow rectangular cells run parallel with the stem and overlap each other at their tapered ends. Sap transfers between these cells through gaps, or pits, that form on one side of a cell and match up with the pits of an adjoining cell. Resin canals are also found in most softwoods. These run alongside and between the tracheids and connect with the horizontal rays that are intended to feed resin to external wounds. These resin-carrying rays are up to 30 cells wide and constitute, on average, 17 percent of the wood volume. All the other softwood rays are one cell thick and make up from 5 to 7 percent of the volume.

Two significant characteristics of softwood are its straight grain, making the wood ideal as light construction lumber, and its long fibers, which are valued in paper manufacturing. Softwood's rays are also aligned, forming straight, radiating spokes. Hardwood grain is seldom as straight, and its lumber is used only where strength, durability, and beauty are overriding factors. The rays of hardwood, like its grain, also are not normally aligned. This occurs because, unlike softwood, hardwood is composed of more than one prominent type of cell, and distortion arises when one type of cell structure pushes another out of alignment.

Hardwoods have four different kinds of cells, whose relative proportions vary with each species. Two of these are storage cells called parenchyma, which hold gums, starches, sugars, and other nutrients. (Softwoods have parenchyma cells as well, but they constitute only a small fraction of the wood.) Elongated cells also develop in hardwood and act solely as strengthening members; these are called fibers. Fibers, together with the fourth type of hardwood cell, the vessel, provide the same sort of structural support that tracheids alone do for the softwoods. Generally, the greater the proportion of fiber in a wood, the higher its density and strength.

Vessel elements are the most distinguishing cell structures in hardwoods. Only rarely occurring in softwoods, they make up between 20 to 60 percent of a hardwood's volume. Constructed of very large cells whose matching ends dissolve to form numerous pipelines, vessels, as with the tracheids of softwood, function to transport sap. Because the diameter of a vessel is substantial (even enlarging as the tree ages), sap is able to flow quite easily through the tree—consider the amount of spring sap that is delivered by the maple tree for making syrup. Up to two-thirds of a hardwood's weight may come from its water content, whereas water accounts for no more than one-third of a softwood's weight. Unfortunately, this higher water content leaves it more susceptible to shrinking when it dries, a bane to woodworkers. Vessels are also responsible for making hardwoods the most beautiful of woods. Their varied size and diverse alignments endow hardwood with a variety of textures, patterns, and luster properties.

As a group, softwoods are less resilient to mechanical injury than are the hardwoods, and because of their high resin content and lower percentage of water, they are more susceptible to forest fires. Their mechanical properties, such as compression strength, tensile strength, impact bending, and, of course, hardness, are also second to the hardwoods. Yet endowed with characteristic lightweight and straight grain, they will always remain an important construction material.

Heath ● Heather

Originally, the word "heath" was used to denote any of the small plants in northern and western Europe that grew on moorland, open woodland, or poor soil. It derives from the German *heide,* the Anglo-Saxon *haeth,* and the Middle English *heth,* for "wasteland." "Heath" in its Greek form, *ereikh,* was eventually used as a basis for the term "Ericaceae," the family designation for some of these plants, and *Erica,* one of its genera.

Plants of the Ericaceae family are most often woody, dwarf evergreen shrubs that, depending on growing conditions, make up communities ranging in height from less than a foot to those over six feet. Along with shrubs, the family of 70 genera and 1,900 species also includes herbs, small trees, and a few vines.

The genus *Erica,* with about 500 species, is popularly called the "heath" genus, and its members are correctly termed "heaths." But other genera of Ericaceae contain plants bearing the name "heath" as well: *Daboecia* holds Saint Daboec's heath, and the *Phyllodoce* genus contains the mountain heath. Yet not all members of Ericaceae, such as the rhododendrons, azaleas, and American laurel, are considered "heaths," which makes the designation "heath family" one of convenience, not a strict definition. Adding to the confusion is the fact that "heath" is also used for a few plants belonging to other families. But, in main, the term usually refers to any of the compact, slow-growing evergreen shrubs and dwarfed trees of the *Erica* genus that are native to Europe and the Mediterranean region.

When the term "heather" is used alone, it invariably refers to one particular species of the heath family, *Calluna vulgars. Calluna,* which comes from the Greek meaning "to sweep," was chosen because brooms were made from its branches, and *vulgars,* meaning "common," indicates its widespread distribution.

Heather, also called ling, is a small, low-growing shrub of Europe and western Asia that generally reaches no more than three feet in height. Once used for bedding, fuel, baskets, and orange dye, it is now primarily managed only as grouse forage. Heather is also called "broom," a name it shares with a plant of the *Cytisus* genus, so misidentification could occur when this term is encountered.

When "heather" is used with a qualifying name, such as bell or purple, it indicates a plant of another genus in the heath family. Thus, beach heather belongs to the *Hudsonia* genus, Himalayan heather to *Cassiope,* and purple heather to *Erica.* So, as with the term "heath," "heather" is likewise used a bit freely within the Ericaceae family, which may in part be a result of common heather's previous classification in *Erica.*

Unfortunately, there is more confusion. Not only are plants in other genera called "heathers," but some of these same plants may also be called "heaths." This is particularly true in the *Erica* genus. The following from *Erica* illustrates this problematic situation.

Species	Common Name		
E. carneal	Snow heather	or	Spring heath
E. cinerea	Bell heather	or	Twisted heath
E. hyemalis	White winter heather	or	French heath
E. tetralix	Bog heather	or	Crossed-leaved heath

Moreover, in the British Isles, "bell heather" also is used for *E. tetralix,* whereas in the United States "bell heather" describes those of *Cassiope mertensiana.* The forgoing illustrates why science readily embraces the Latin binomial method of naming plants, animals, and other organisms.

Mold • Mildew

For the most part, molds are an unwanted intrusion in our lives. Not only do they absorb nutrients from both dead and living tissue, rendering our foods inedible and our plants weakened, but they often discolor and mar the surfaces on which they grow.

Most belong to the Fungi kingdom, where they comprise about 600 species. Fungi, as a whole, are primarily terrestrial organisms made up of a curious assemblage of growths that botanists have organized into three general categories: those forming filaments that crowd together to shape large fruiting bodies, such as mushrooms, truffles, and lichens; those forming a single spherical cell, the yeast organisms; and those whose multicelled branching filaments remain ununited, such as the molds. Because under certain environmental conditions some fungi species can occur as either a mold or a yeast, neither of these terms is used to formally categorize fungi. Furthermore, not all organisms with "mold" in their name are fungi. Slime molds and water molds, more primitive organisms that were once members of the Fungi kingdom, have been reassigned to the kingdoms Myxomycota and Oomycota, respectively. (Both of these kingdoms are new and somewhat tentative. See "Kingdom—Species".)

The basic structural unit of fungal mold is the hypha, a long thread-like cell or a chain of cells. As these hyphae develop, they branch out and interconnect to form an assemblage of vegetation called a mycelium, which eventually grows large enough to be visible to the naked eye. This mass of conspicuous interwoven filaments with fruiting structures is the fungal body we call mold. Rusts and smuts, other fungal diseases of vegetation, are similar in appearance to molds, but because of their unique reproductive bodies, each belongs to a different class of fungi.

Fungal molds, only a fraction of an inch high, appear as fuzzy or woolly growths on living and nonliving bodies. Some, such as the sooty mold, attack growing plants, whereas others live as parasites on insects or other fungi. But most molds exist on nonliving organic matter. Our food is an excellent medium for their growth, whether in a natural form like fruit or meat or a processed product such as bread, cheese, or canned fruits and jellies. Other materials appealing to molds include paper, leather, linen, and fabrics.

All molds are potentially destructive organisms, invading their host and breaking down its cellular structure. Yet a few are beneficial to man. Bread molds are used to make sufu, a cheese from China; tempeh, a soybean food of Indonesia; and, most important, penicillin. The streaks of dark blue in blue cheese are also a mold.

As mentioned, besides the fungal molds, there are also slime molds and water molds. Slime molds come in two distinct forms, neither of which forms anything like a wooly growth. The plasmodial slime molds (phylum, Myxomycota) lack a cell wall and travel over a surface feeding on bacteria, fungal spores, and small bits of plant and animal material. They form a very thin body that may cover several square feet, often on logs, cacti, or the soil. Most become conspicuous when they dry out and turn bright yellow or orange. Cellular slime molds (phylum, Dictyosteliomycota), like fungi, have cell walls, although these walls are composed of cellulose, rather than the chitin of fungal mold. More like amoebas than anything else, individual slime mold slugs will migrate together to form an aggregation that creates a structure composed of base, stalk, and spore-bearing mass on top. Because of their structures and lack of fuzzy appearance, neither plasmodial nor cellular slime molds are apt to be confused with the fungal molds, something that cannot be said of the water molds.

Although water molds usually exist on underwater surfaces, they still present a wooly or cottony appearance. Once thought to be fungi because of their filamentous growth and similar feeding habits, they are now set apart from the fungal molds because their cell walls are composed of a mix of cellulose and glycan, rather than of chitin. Furthermore, the nuclei within their filaments each have two sets of chromosomes, rather than the single set that is characteristic of fungal molds. Also, unlike the fungal mold, their cells have flagella, whiplike structures that enable them to swim in moist environments, principally water. Although most are not particularly harmful, some do cause serious diseases in fish.

Because these organisms make up the bulk of the Oomycota's more than 500 species, the term "water mold" is commonly used to encompass all its members; however, a few in the kingdom are more often designated as mildews. Called "downy mildews" because of their fuzzy, downy appearance, these fungi-like growths have the same defining characteristics as the water molds, except that they are terrestrial. Comprising the family Peronosporales, the family's 15 species typically appear on the underside of leaves, forming grayish, downy patches, or on top of leaves as yellow spots. They also will attack the fruit of a plant where moisture is easily accessed. Many species are true parasites that live not only on a plant's surface but in its tissue as well. Grapes are particularly favored, as are other fleshy fruits. More often, however, it is the leaves that are infected, including those of lettuce, onions, tobacco, clovers, cucumbers, corn, sugar cane, and cabbages. But these are not the only mildews to attack plants. Another mildew form, called the "powdery mildews," is just as infectious.

Powdery mildews get their name from their soft, powdery appearance, which, although very darkly colored in some tropical species, is typically white or off-white. These mildews do not invade the tissue of their host plant, as do the downy mildews, but lie on its surface and with probing suckers draw nutrients off the top layer of cells. More important, however, powdery mildews are not even remotely related to the downy mildews but are members of the Fungi kingdom.

Only somewhat akin to the fungi molds (fungi molds belong to the phylum Zygomycota and powdery mildews to the much larger phylum Ascomycota), their spores, unlike those of mold, are not borne in specialized hollow structures of their filaments but develop in distinct sac-like structures. Furthermore, their methods of reproduction are vastly different.

Powdery mildews also differ in their lifestyle as obligate parasites—unlike the molds, they can grow only on living plants. Although their mode of feeding off the surface cells of vegetation may not seem particularly harmful, the health of plants such as grass, corn, wheat, roses, dandelions, apples, and clover can be seriously compromised—only rarely will they kill a plant. Stunted growth will sometimes occur, but more often a plant's leaves will tend to dry and shrivel. As a whole, a powdery mildew will sap the plant of nutrients and decrease its ability to photosynthesize.

The terms "mold" and "mildew" do not have very precise meanings. "Mold" encompasses the fungi molds of Zygomycota, the slime molds of Myxomycota and Dictyosteliomycota, and water molds of Oomycota. "Mildew" covers not only the powdery fungi mildews of Ascomycota but the downy "water" mildews of Oomycota as well. Although one could well disregard the two slime molds—neither has the fuzzy appearance typical of any of the molds or the mildews—the others lack any defining characteristic that can be said to be typical of just molds or just mildews.

Moss ● Lichen

Next to ferns, mosses are the most advanced of the spore-bearing plants. Because they lack a true vascular system of food- and water-conducting tissue and have no coiling mechanism with which to disperse spores, they are placed in a division of primitive land plants called Bryophyta. Bryophytes are green plants that are recognized for their reproductive cycle of alternating generations (liverworts and hornworts also reproduce in much the same way). The egg and sperm generation produces the green, leafy photosynthesizing body, whereas its complement, the spore-bearing generation, forms a much smaller structure, which relies on the first generation for nutrition (see "Spore").

Structurally, moss is mostly water. Because it has such a poorly developed system for transporting fluids, it must store all the water it can in order to weather even moderate dry spells. If a dry spell lasts too long, maintaining a proper water balance is impossible and the moss dries out. Deprived of water, it will go dormant for periods that in experimental cases have lasted up to ten years. This adaptation to water deprivation is called poikilohydry. During this time, the "leaves" of the moss curl up and become brown and crusty. Given enough water, however, the plant quickly revives, begins photosynthesizing again, and resumes its regeneration cycle.

Most mosses are small, low-growing plants that, unlike more advanced vegetation, do not have true roots, stems, or leaves. They inhabit humid meadows, woodlands, and wetlands, where they can be found on tree trunks, on fallen branches, on rocks, and, most significantly, in bogs. In fact, sphagnum moss constitutes the major component of most bogs. Accumulating over the years on still waters, sphagnum moss eventually turns into peat, a compacted, partially decomposed turf that in the past was used as a low-grade fuel. Given thousands of years and the proper conditions, peat will eventually transform into a vastly better fuel: coal.

In the humid woodlands of the North American Pacific coast, blankets of moss can be found enrobing tree trunks and rocks, while other species hang from branches in curtains that resemble the Spanish moss of the U.S. Southeast. However, not all "moss" is as it appears. Although called a "moss," Spanish moss is really a member of the pineapple family, and Irish moss, which is often found on the seacoasts of the north, is an alga. The same can be said of reindeer moss, the principal food of reindeer and caribou, which is really a lichen.

Lichens, a few of which resemble the draped mosses of trees, are most often mistaken for mosses when found on rocks and bare ground—just about the same places many mosses grow. Lichens also grow in environments hostile to other plant life. They have been found on blistering desert rock, in the Arctic and the Antarctic at below-zero temperatures, and high in the mountains just below the snow line. However, one of the few environmental conditions they cannot tolerate is polluted air; therefore, seldom will they be found in large industrial cities.

Considering the habitats and the appearance of lichens, it is not surprising that they are often mistaken for mosses. Only on close examination can the lichen's particular character be discerned, exposing it as one of the most unique plants on Earth. Oddly enough, it is not so much one plant as it is two in partnership—one a fungus and the other most often an alga. In some instances the fungal partner will be a cyanobacteria or a xanthophyte (a protoctist). Because the fungal element is its major component, lichens are placed in the Fungi kingdom and named, taxonomically, for their fungal component. At present, about 13,250 species of lichen fungi have been identified. Although they are classified as fungi, this does not mean that the algal or the cyanobacterial partner plays a lesser role in the lichen's life. These partners lie between the outer layers of fungi or are spread throughout it and, through photosynthesis, provide all the carbohydrate and nitrogen requirements for the fungus. In return, the fungus provides a structure in which the algae can live and nourishes the algae with the nutrients it absorbs from the air and its substrate.

From studies made in 1981 of the lichen known as British soldiers, it was found that this partnership may not always be as equal as first thought. In this particular lichen, algae cells were destroyed as they nourished the fungus, and it was only by the continued replication of the yet-untouched cells that the algae population was able to keep up with its mortality rate.

Although some lichens encapsulate more than one kind of algae, most harbor just one species. In contrast, the fungal component is always a solitary species unique to a specific lichen. Some algae can exist apart from their fungal partner, but it is rare to find a lichen fungus growing alone.

Based on their appearance, lichens fall into one of three categories: the crusty-appearing crutose lichen, the leaflike foliose lichen, and the fructicose lichen that resembles Spanish moss or tiny bushes. Regardless of their form, they all grow quite slowly, the fastest gaining only 0.4 of an inch a year.

Reproduction is primarily by asexual means, either by casting off large pieces or by dispersing very small bits of itself called soredia. Sexual reproduction can occur within the fungal component, but a new lichen can form only if a fungal spore can germinate near and capture an algal partner.

Mushroom ● Toadstool

Probably more words have been expended over the distinction between "mushroom" and "toadstool" than over any other two terms in this book. The crux of the problem is that neither word is a scientific designation. Both have gained acceptance as popular, culturally influenced terms whose spellings and definitions have continued to shift throughout the years.

The term "mushroom" comes from the Old Norman-French *mousseron,* which in turn came from either the French *mousse* or the Norse-German *moos,* both being broad terms for moss. Today *mousseron* is still used in France to denote exceptionally good mushrooms—in particular, St. George's agaric. "Toadstool" may also have come from the German language. It is said that because of the toxicity of some stem-and-cap varieties, the Germans called them *todstuhl* or *todesstuhl,* meaning, "death stool." Further speculation suggests that the British borrowed the word and, because of its sound, wryly changed the meaning to "toadstool," a stool for toads. In another, more likely version of its origin, "toadstool" is thought to have come from the Old English word *tadestole* which is comprised of *tadde* (toad) and *stole* (stool). The *tadde* is suggested because both the fungus and the amphibian can be somewhat poisonous, and they inhabit shady, dank, secluded places. This derivation is supported by the numerous "toadstool" words found in other cultures around the world—many of these cultures being not only geographically but also linguistically isolated from one another. But whatever their language, people throughout the world seem to have regarded the fungus in an almost identical manner: making some kind of connection between it and toads.

Today, unfortunately, the term "mushroom" still has no definitive meaning, with "toadstool" faring only somewhat better. To illustrate, consider the following conflicting statements taken from various sources published over the last thirty-some years.

Are all mushrooms edible?

> "The term 'mushroom,' as commonly used, narrows the field to those fungi that can be eaten."[4]

> "Mushrooms: the fruiting body, of a fleshy nature, characteristic of some fungi. It is applied to both edible and poisonous species."[5]

Are morels (an edible fungus) mushrooms?

> "Morels and truffles. Neither is a true mushroom."[6]

> "Sponge Mushrooms: Morels."[7]

Are fungi such as puffballs, bracket fungi, and jelly fungi, mushrooms?

> "Are puffballs mushrooms? Yes, they are mushrooms."[8]

> "Mushrooms without gills: puffballs, morels, jelly fungi, bracket fungi."[9]

> "Other members of this class [Basidiomycetes] include mushrooms or toadstools, puffballs, earth stars, shelf or bracket fungi, jelly fungi . . ."[10]

As for toadstools, whose etymology implies a less-than-wholesome origin, it is fair to assert that they are mushrooms of some sort, are possibly parasol-shaped and, if not poisonous, are at least inedible. This is, in fact, the most popular view of the toadstool. Not surprisingly, there are other opinions:

> "The only distinctions between mushrooms and toadstools are based on folklore or tradition—botanically there is no difference."[11]

"It is common for certain fungi to be recognized in some areas as mushrooms and considered edible, while others are termed toadstools and believed poisonous. In other places the terms are used synonymously without regard for supposed toxicity."[12]

To confuse the issue a bit more, in Great Britain the term "toadstool" most often covers all fruiting fungi, with "mushroom" used to denote only those toadstools used as food—in particular, the common edible field or meadow mushroom. As biologist Gary Rahan put it, "In Great Britain the only fungi that qualifies as a mushroom is the commonly cultivated species you find in the supermarkets."[13] This is somewhat the opposite of the prevailing North American view that "mushroom" covers, to whatever extent, fruiting fungi, and that toadstools are more or less poisonous varieties of mushrooms.

Considering the various opinions on what constitutes a "mushroom" and a "toadstool," the popular notion that mushrooms are stem-and-cap fungi and that toadstools are only poisonous varieties of mushrooms is as valid a view as any. This, of course, is an arbitrary choice of definition, as are those definitions found in mushroom guidebooks—definitions that should be well understood when used in identifying wild fungi.

Muskmelon ● Cantaloupe

For years "muskmelon" served as an alias for a number of different melons. The crenshaw, the honeydew, the casaba, the cantaloupe, the Persian, and the muskmelon, due to their common classification as *Cucumis melo,* have all been called a "muskmelon" at one time or another. But ever since botanists split up the six melons into three groups, "muskmelon" has become a more restricted term. These three groups are: *C. m. reticulatus,* the Persian and the muskmelon; *C. m. inodorus,* the casaba, the crenshaw, and the honeydew; and *C. m. cantalupensis,* the cantaloupe. All members of *C. melo* are warm to temperate annuals, with soft, hairy, rigid vines that trail along the ground or climb supports. The family of these melons, Cucurbitaceae, also includes the cucumber and squash.

Most popular varieties of muskmelons are spherical, are slightly segmented with broad ribbing, and have a surface covered with a raised complex of netting that is usually lighter than the skin underneath. Less typical varieties have shapes approaching that of the banana and rinds that may have little or almost no netting. Inside the melon is a soft flesh that ranges from white to pale green to orange. What sets many of the muskmelons apart from the other melons, and is the source of their name, is the distinct musky scent they emit, which in turn is why some of those with very little scent are in a group called *inodorus* (without odor).

The cantaloupe, *Cucumis melo cantalupensis,* does not have a netted rind like the muskmelon but rather a rough warty or scaly covering that is often grooved and green. This hard-rind melon has flesh that tastes much like the muskmelon's, but its color is mainly orange or green, never white. Named after the former Italian papal estate of Cantelupo, it is the most widely grown melon in Europe. Principally a Mediterranean fruit, it is hardy enough to grow in Britain, whose climate is considerably cooler than that of its place of origin in Southwest Asia. Introduced to North America during the colonial period, it was a popular melon for some time but eventually fell out of favor. Today, the cantaloupe is rarely grown in the United States.

But what of all those cantaloupes sold in grocery stores and at vegetable stands? The boxes they come shipped in are frequently labeled "cantaloupe," and major seed distributors—who should certainly know a melon when they see one—sell a variety of seeds for growing cantaloupes. If these are not cantaloupes, then what are they? They are, in fact, all muskmelons: *Cucumis melo reticulatus.*

Most commercial melons sold in the United States come from the West Coast and the South, where their growers label the smaller round muskmelons that ripen slowly "cantaloupes." These fruits have tough, hard rinds with only slight ribbing and stand up well to shipping. The larger varieties that ripen quickly, develop a thinner skin, and do not ship as well usually remain identified as muskmelons. Grocers, looking for melons that arrive unbruised, invariably select the mild-flavored "cantaloupe" type (also called a western shipping melon by its growers) and sell it under the name "cantaloupe"—erroneous as it is. However, not all commercial muskmelons are misidentified; a few, characterized by a more oblong shape, with broad-ribbed and coarse-netted rinds, are more often sold under their rightful name in eastern and local markets. These larger melons have a deeper orange-colored flesh, with a sweeter and more aromatic flavor than the "cantaloupe." However, as distinct as they may be, even these varieties are being mislabeled "cantaloupes" more and more frequently.

Why growers, grocers, and seed producers prefer to call muskmelons "cantaloupes" is unclear, but the practice appears quite widespread. In fact, many popular gardening and produce reference books use the term "muskmelon" only parenthetically, if at all. One source even misidentifies the muskmelon as belonging to the cantaloupe group *cantalupensis.*[14] To find a true cantaloupe in a North American market would be unusual indeed. Unfortunately, the term "muskmelon" seems to be on the way out, just as "sweet potato" appears likely to be eventually replaced by "yam."

Onion ● Scallion ● Shallot ● Leek

Most onions are biennial plants that store food within their bulbs during their first year of growth and then flower in their second year. Their round leaves and hollow stems grow about 12 inches high in the bunching forms and up to 20 inches in the single-bulb varieties. Both types, native to western Asia, belong to the *cepa* group of *Allium cepa.* Their underground bulbs, which grow close to the surface, range in diameter from almost nonexistent in the bunching types, to between 1 and 5 inches in the bulb onion. Unlike garlic, another bulb of the *Allium* genus, onion bulbs do not form cloves but develop as a single structure of concentric layers of leaves.

The bunching onion, sometimes called a green onion, gets its name from the way its tubular stem divides at the base and sprouts into a cluster. These onions, usually white or green, are often harvested before they mature and typically will be eaten raw. Bulb onions, usually more pungent than bunching onions, are just as often cooked or fried. Unlike the bunching varieties, the casing of the bulb onion develops a dry, papery skin, which varies in color from silver white to yellow to a dark brown or red.

Although most bulb onions have a fairly round shape, some develop a distinctly oval or pronounced flattened form. Their flavor, as with all onions, is strictly dependent on the concentration of sulfur compounds, which also gives them their pungent aroma.

"Scallion" is a term that suffers considerable abuse in application. It is not uncommon to find lay people, as well as some authorities, using "scallions" for onions, young leeks, and shallots. Even unabridged dictionaries list these three plants as common definitions or synonyms for the word "scallion." Some writers contend that the scallion is a distinct plant, *Allium fistulosum,* a species that also includes the Japanese bunching onion, the Welsh onion, the ciboule (cibol), and the Spanish onion. However, botanical reference works seldom mention "scallion" in conjunction with *Allium fistulosum.* In fact, the term is not recognized as a botanical designation at all.

While scallions were originally specific green bunching onions from the city of Ascalon in Palestine, the term has evolved to denote any bunching onion or the young shoot of a bulb onion that is picked before maturity. The use of "scallion" for a leek or a shallot is questionable unless the plant is harvested before maturity, which is not a common practice. So "scallion," properly used, may denote a number of species, providing such plant is fairly young and its most distinguishing characteristic is its lack of a bulb—although slight swelling may be present.

Shallots, once considered a distinct species *(Allium ascalonicum),* are now classified with onions in *A. cepa* but in their own group, *aggeratum,* which also includes the potato onion. Sometimes called eschallots, they are thought to have originally come from western Asia and been distributed throughout Europe by the Crusaders. The slender stalks of young shallots closely resemble those of bunching onions and are sometimes sold as such. (This is also true of certain onions not of *A. cepa* that resemble the green bunching onion.)

However, shallots can usually be distinguished by their grayish bulbs, which tend to be angular and pointed. In taste, they are considerably milder and more delicate than the onion. But the most distinctive feature of the shallot is its bulb, which, like the garlic, develops cloves. Each clove is composed of multiple layers with a parchment-like covering, similar to that of onions. It is through propagating these cloves that the shallot usually reproduces, because, unlike the onion, which can be grown by seed, the shallot is often sterile and therefore seedless.

Like the onion and the shallot of *A. cepa,* leeks also belong to the "onion" genus *Allium,* but because of several unique characteristics they are placed in the *porrum* group of *Allium ampeloprasum.* Cultivated and bred for over 4,000 years from various species, the leek is no longer a truly wild plant. Unlike the onion, it does not develop an actual bulb but grows a thick white cylindrical underground stem up to 2 inches in diameter and from 6 to 18 inches long. Above ground, its rather tough straplike leaves grow 2 to 3 feet high. Usually grown from seed, the leek is cultivated as a biennial. If the plant is not harvested but is let to go to seed, a corm forms at the base of the stalk, which can then be replanted to produce a new leek the following year. Neither onions nor shallots bear corms. The fragrance of a growing leek is much like that of an onion, but having a milder and sweeter flavor, it is just as likely to be eaten raw as cooked.

Pine ● Fir ● Spruce

As a matter of convenience, we frequently call the conifer forests of North America "pine forests," regardless of the species of trees they may contain.

"Pine" is also freely used for smaller stands of needled trees and even individual trees, all of which may actually be firs, spruces, hemlocks, tamaracks, and so on. Europeans are equally guilty of lumping the conifers together, but instead of using "pine," they often use the collective term "fir." The misapplication of these labels is not usually done out of ignorance but because the two terms function as a convenient shorthand. Most people, even if unacquainted with the differences between the various trees, recognize that pines, firs, and spruces are distinct conifers. These trees, in fact, have such individually unique features that each is accorded its own genus.

Unfortunately, explorers and settlers of the New World, ignorant of specific species' characteristics, misnamed a number of conifers. And scientists, in reevaluating plant relationships, have changed the classifications of some. Consequently, we have trees like the Douglas fir, the Chinese fir, and the big cone spruce, which are not true firs or spruces. Also, several so-called pines belong to genera other than the pine genus *Pinus,* such as the black pine, the brown pine, the Norfolk Island pine, the King William pine, the Japanese umbrella pine, and the Tasmanian Huon pine. Custom has preserved these misnomers, but the clues for identifying the trees of each genus are very apparent, so the proper identification of a tree, whether misnamed or not, is not difficult.

Although all pines, firs, and spruces are distributed extensively throughout the northern hemisphere, it is the pine that predominates. Of *Pinus'* 90-plus species (the largest genus of the family Pinaceae), 40 are native to North America. Pines extend from Nicaragua to the tree line in Canada and also flourish in the West Indies and Eurasia. Forming extensive forests, they are most valued as a source of timber and pulpwood. The logging industriy has prompted considerable cultivation and reforestation because of overcutting in past years.

Some 40 species of fir trees make up the genus *Abies,* 11 of which can be found in the United States. These tall, straight, and narrow conifers are generally pyramidal, with branches so dense that young firs appear almost perfectly conical. Older specimens, which often attain heights over 200 feet, eventually lose branches and become a bit more ragged looking in their later years. Unlike the other conifers, firs do not shed their needles after being cut down and therefore are very popular as Christmas trees. But their principal value is in providing wood pulp for paper manufacturing and as light construction material.

Unfortunately, popular nature books often place the Douglas fir in with the true firs. Although once classified as a fir, it is now a member of the genus *Pseudotsuga,* which translates as "false hemlock": a fitting description, in that it is not a true hemlock. Because it has retained its misnomer, the Douglas fir (also called a Douglas spruce and a red fir) is apt to leave an erroneous picture of the real firs. The Douglas fir, known for its great height—some reaching 300 feet—is second only to the redwoods and the giant sequoia in size. A valuable tree because of its lumber, the Douglas fir grows primarily along the coasts of Washington and Oregon.

Spruce trees belong to the genus *Picea,* a name derived from the Latin word *pix,* for "pitch." A common tree in the northern hemisphere, the spruce grows best in cool, moist regions. Compared to the pine and the fir, spruces are much shorter trees; the Norway spruce of Europe, the tallest species, reaches heights of only 200 feet under the best growing conditions. Yet spruces exceed fir trees in population and number of growing regions. Of the 40-some species worldwide, 30 grow in the United States. Some spruce trees are cut for lumber, but more often they end up as pulpwood.

A number of variations in the features of the three trees makes each genus unique. There are too many variations to list here; the most significant ones, however, those that will aid in a quick identification, include:

Bark

Pine	Furrowed and scaly
Fir	Smooth but broken by blisters of resin
Spruce	Scaling and flaking, never ridged

Branches

Pine	Most in a whorled pattern on the trunk
Fir	In a whorled pattern, usually held horizontally
Spruce	In a whorled pattern, often bent downward

Cones

Pine	Large, round, and woody. They occur singly or a few in a cluster throughout the tree. Some scales are prickle tipped.
Fir	Long, erect on branches, mainly in top branches. Scales fall off while cone is on tree, leaving a spike by winter. Scales are

| | never prickle tipped but rather are round topped. |
| Spruce | Ovoid or cylindrical. Unlike the fir, they hang down and remain on the tree for 2 to 3 years. Scales have no prickles and resemble a corn flake; found throughout the tree. |

Needles

Pine	Long, stiff, and round or triangular. They vary in length from 1 to 12 inches. Persist on branches for 2 to 3 years. Occur in bundles of 2, 3, or 5, rarely singly. Base of bundle wrapped in a parchment-like sheath.
Fir	Short, soft, and flattened, often with a blunt end. Usually green, with two whitish bands, and grooved on top. Underside often silvery. Arranged as if in two ranks. When they drop off, they take their attaching base with them, leaving a smooth branchlet. Persistent when dry but easily pulled off. No sheath.
Spruce	Short, ridged, and often spine tipped. Most are generally four-sided in cross-section. Color ranges from blue to green. Arranged in a spiral on the twig, they leave short pegs when removed, resulting in rough branchlets. Although they shed easily when dry, on a green tree they adhere better than the fir needle. No sheath.

Redwood ● Sequoia

Of all the trees in the world, none are more awe-inspiring than the redwood and the sequoia. Incredibly tall and massive, they are some of the oldest living organisms on Earth; one, since cut down, was estimated to have been over 4,000 years old. Native to central and northern California and a small portion of southwest Oregon, they are frequently mistaken for each other or are even thought to be one and the same tree. Because of their close kinship, they were once treated as different species in the *Sequoia* genus; the sequoia has only recently been given its own genus, *Sequoiadendron*. Now the only member of the *Sequoia* genus is *S. sempervirens*—the redwood, also called the California redwood or the coast redwood. The lone *Sequoiadendron* species is *gigantea,* whose other popular names include bigtree, giant sequoia, and Sierra redwood (an unfortunately misleading name). Both trees date back to the Cretaceous period, when their ancestral forms ranged as far north as the arctic realm.

Today, redwoods are restricted to the northern coastline of California and the southern margin of Oregon in an intermittent run of trees 500 miles long and from 5 to a little more than 30 miles wide. Here they thrive on the moisture-laden air swept in from the Pacific Ocean. Abundant water, plus a cool climate, has encouraged them to become the tallest trees in the world, some reaching close to 370 feet in height. Growing best on protected flat land near streams, deltas, or tracts bordering the Pacific, they are often found growing among firs, cedars, oaks, spruces, and hemlocks.

As a conifer, the redwood is unique in its reproduction; it not only grows from seed but also sends up shoots from roots and stumps. This is very advantageous, in that the seeds themselves are poor germinators. Because young shoots can sprout from lateral roots, young trees will sometimes closely encircle stumps. Competition among these saplings eventually eliminates those less vigorous or hardy. Although the redwood is named for its heartwood, which varies in shade from light to dark red, its sapwood is quite light, not unlike pine wood. The redwood has long been a favored construction wood; its durability and resistance to decay and insects make it ideal for use outdoors, where it is in direct contact with the weather and the soil. Better-looking pieces are used to make veneers, decorative works, and furniture.

Between 1770 and 1843, a famous Cherokee Indian "half-breed" (born in Tennessee of a German mother and a Cherokee father) named Sequoyah traveled west to live and teach. He was revered among the Indians as a wise and fair man who, among other things, invented the Cherokee alphabet. It was after him and in his honor that the sequoia was named.

The sequoia, unlike the coastal redwood, is an inland tree that dots the western side of the Sierra Nevada mountains in small isolated stands of up to 1,000 trees. Growing at elevations from 4,000 to 8,500 feet, where most of the precipitation arrives as snow, these trees enjoy a much colder and less humid climate than do the redwoods, which rarely grow above an altitude of 3,000 feet. Although not reaching the heights of the redwood, the sequoia, which can grow up to 311 feet high, is a much more massive tree, often touted as the largest of all living things in the world. The greatest sequoia ever measured had a girth of 116 feet, whereas the largest redwood reached only 62 feet around. Sequoia do not usually grow in close association with other species but prefer to establish purer, more sparsely populated stands. Those trees that do occasionally surround clusters of sequoia include ponderosa and Jeffrey pines, red firs and white firs, and the incense cedar.

The wood of the sequoia is lighter in weight and more brittle than that of the redwood. This, plus the ungainly size of its trunk, made the sequoia a runner up to the redwood in the logging industry. Yet sequoia were still overcut, which so reduced their numbers that today the resulting supply and market for sequoia timber have made their logging unprofitable.

Along with their different growing conditions and timber characteristics, the following are some of their more apparent physical differences.

Feature	Redwood	Sequoia
Needles	½–1 inch long, linear	¼–½ inch long, more oval
	Ranked in twos on twig	Thickly covering twigs
	Dark yellow-green on top	Blue-green, turning brown
Cones	15–20 scales	25–40 scales
	Mature in first year	Mature in second year
	¾–1 inch long	2–3½ inches long
Twig	Smooth	Covered with needlelike scales
Bark	6–12 inches thick	12–31 inches thick
	Closely furrowed	Furrows between wide ridges
Crown	Full and conical	Scrawny

Sap ● Resin ● Pitch ● Gum

Sap is the lifeblood of trees. It circulates from the roots up through the trunk to the leaves and then back down to the roots. Similar to the way that blood travels, each of its two trips—up and down—takes a different pathway and transports different life-sustaining nutrients. Sap forms in the roots from groundwater; cell walls filter the water of bacteria and other organic matter, leaving it almost pure. What is not filtered out are the forty-some elements taken from the soil that the tree needs for growth, such as copper, iodine, sulfur, iron, calcium, carbon, and phosphorus. Sodium chloride (salt) and other undesirable chemicals that may have also been drawn in are eventually dissipated through evaporation in the leaves. To reach every leaf and bud, the sap flows through a ring of wood called, appropriately, the sapwood, or xylem, which lies between the heartwood (if it has formed) and the thin, growing layer of wood, the cambium. The amount of sap circulating in a tree is tremendous. At least half the weight of a log is water, and growing leaves, twigs, and fruit are over 80 percent water. Some trees, such as the willow, absorb up to 50 gallons of water a day. It's been estimated that in one day the average tree will circulate close to 300 gallons of water, almost all of which eventually evaporates. In fact, the tree retains only 1 percent of the water in sap.

Within the leaves, a portion of the remaining water combines with absorbed carbon dioxide to produce carbohydrates of sugar and starch—the process of photosynthesis. These nutrients, together with proteins, oils, fats, tannins, vitamins, and other substances, all in a water solution, constitute the second form of sap that nourishes cell growth and provides energy to the roots. This sap travels down the tree to the roots through a spongy layer of inner bark called the phloem, which lies between the outer bark and the cambium.

With the onset of winter's cold, sap circulation comes to a halt. During this period of dormancy, the sugar within the wood of some tree species dissolves into the sap. In addition, in the warmth of spring, stored starch in the trunk and the roots converts to sugar, which is transferred to the xylem, where it combines with root sap. This sap is then conducted up the tree, where, with the help of hormones produced in the roots, it prompts bud growth. This is the sap in maple trees, from which syrup is made. It must be noted that although trees have been the only plants discussed, almost all green plants manufacture sap.

Resins, unlike sap, do not help nourish a plant but are compounds secreted in response to injury. A variety of plants produces this complex substance, which is exuded at an injury site to ward off insect invasion and disease organisms. The most familiar resin, called oleoresin, is that of the pines. Other resins are the hard resins, including those from trees of the legume family; damar resins from several trees found in southern Asia and the East Indies; and gum resins, a combination of resin and gum that is produced in such plants as frankincense and myrrh.

Resins are complex compounds of carbon, hydrogen, and oxygen that easily separate into two principal components: a solvent (the fluid medium) and a solute (particulate matter, which remains after the solvent evaporates). In the pine family this solvent is turpentine, a volatile oil. The remaining ingredient, the solute, is a waxy substance known as rosin. When a plant secretes a resin, it forms a protective cover over the injured area. Exposed to the air, the volatile oils gradually evaporate, leaving behind a hard, translucent substance. Given enough time, this will chemically change to fossil amber, or copal if from a hard resin. This is a significant point because so often people mistake a tree's sap to be the source of amber—an echo of the popular misconception that the sticky resin of a tree is its sap. An important characteristic of resin is its insolubility in water, which provides protection from moisture loss in a wound. It is, however, soluble in alcohol, ether, and other organic solvents.

Most resins accumulate in vertical resin canals as the tree grows, although firs, hemlocks, and the sequoia form resin canals only at the time of injury. These resins are supplied to the canals by other surrounding cells, called parenchyma cells. Both structures feed horizontal resin canals that extend in rays to the bark.

Pitch is the common name given to the pine oleoresins, particularly when they form on the outside of a plant and have not yet dried out. The residual oil, turpentine, keeps it soft and very sticky, a property that led to its use years ago as a coating over the caulking in wooden ships. This pitch should not be confused with the petroleum- or bituminous-based pitches that are obtained by boiling or distillation.

Those trees and other plants, including seaweed, that do not protect themselves with resin use other compounds, such as latex and gum. The familiar latex, from which we derive rubber, is a composition of gum, sugars, proteins, salts, oils, and other substances. Gum itself is a soft, sticky polysaccharide that forms in the intercellular parts of a plant and as an extracellular secretion. A much different chemical mixture than resin, gum is soluble in water and, conversely, not in resin solvents. Used internally, often in seeds, tubers, and roots, it protects germinating seeds and assists in retaining water. Secreted like a resin, gum also acts as a defense at an injury site by shielding the wound or washing away harmful agents such as fungi spores and insect larvae. More often, though, gum flows in response to boring insects.

In trees, gum ducts form in the inner bark and, in some species, in the cambium and the new xylem as well. All exuded gums are somewhat more chemically complex than are internal gums. This may mean that in some plants, they are generated at the injury site rather than somewhere else in the plant—the actual process is uncertain. Regardless of just where in a plant the gum originates, as a defense mechanism it unfortunately exacts a price. The manufacturing of gum demands a great deal of stored energy, and if demand is significantly large or occurs over a long period, it may impact other functions and seriously weaken the plant.

Spine • Thorn • Prickle

Spines, thorns, and prickles are hard, sharply pointed outgrowths on plants, whose main purpose is defense—a function they perform quite well, as most of us can easily attest. Yet some are not without enemies themselves. The acacia plant of the savannas and the steppes has thorns that are particularly vulnerable to ants, which burrow tunnels into the thorn base to excavate a nesting cavity. Man has also found uses for these structures, particularly in the more primitive cultures. Tattooing, skin piercing, and sewing were done using spines and, less commonly, thorns and prickles. The three also made excellent barriers, either as dead brush or live shrubbery for keeping domestic animals corralled and predators at bay—a precursor to today's barbed wire. Less laudable was their use as instruments of humiliation and torture, such as the infamous crown of thorns.

Although their functions and shapes are similar, and despite the common practice of using the three terms interchangeably, each is a unique plant structure. One writer, in describing how the pasture gooseberry has spines and prickles that evolve on the same stem, calls the plant a thorny shrub—thorns being the one pricking form it lacks.

Spines and thorns develop as outgrowths of the inner woody tissue of stems and, as such, are difficult to break off. The spine is actually a modified leaf that does not photosynthesize, or a stipule, a small structure at the base of a leaf stem. Spines on plants like the barberry, the Barbados gooseberry, the black locust, and the Parkinsonia complement the conventional leaves these plants use for photosynthesis. But on the cactus, which has no such conventional leaves, the only form of leaf present is its needlelike spines (photosynthesis occurs in its swollen stem). On the black locust the spines emerge from the stipules on both sides of the leaf stem, whereas on the Parkinsonia, not only the stipules but also the leaf's midrib develops into a spine. Unlike the slender needles of the cactus, these spines more closely resemble the rose "thorn."

Thorns, also called stem-spines, are modified branches or stems that are usually squatter than spines, although in some plants, like the fire-thorn, they tend to be thin and sharply pointed. Somewhat like the stip-ule, they, too, are growths at the base of a leaf but in the axil. The axil is the upper or smaller angle formed by a leaf stem and the branch from which it arises. Unlike spines, some thorns, because they are a form of woody stem, produce leaves along their length. As woody structures, they are very stiff and rugged and, like spines, quite dangerous. Other plants bearing thorns include the honey locust, the osage orange, and the cock-spur hawthorn—but not the rose.

Rose plants, contrary to popular usage, prick us not with thorns but prickles. Termed "emergences," prickles do not arise from interior plant tissue but grow out of the epidermis or bark, and, somewhat like hairs, they do not normally have vascular tissue. Because they grow from the surface sheath and contain no wood, they tend to be somewhat flexible and weak, breaking off much easier than do spines or thorns. However, they can still produce a good puncture. Sometimes growing on the surface of leaves, they are able to protect a plant from even the most aggressive animal. More than protection, however, prickles enable a plant to climb above other ground foliage by fastening into walls, trees, and other veg-etation to reach unshaded sun. One of the most common of these clinging plants is the blackberry, whose long and flexible stems reach lengths of over 15 feet.

Spore • Seed

The evolution and the survival of vegetation have long been underscored by an ongoing struggle against a host of adversities. Hostile geography, adverse weather, competing plant life, invasive insects, hungry herbivores, pathogens (bacteria, parasites, and viruses), as well as we humans, con-tinually challenge a plant's foothold on Earth. Survival demands that none of its functions be seriously compromised—most important, the ability to reproduce. The flora we have today successfully evolved by using a number of reproductive techniques, each plant[15] form adopting methods best suiting its makeup and the obstacles it faced.

Of all these techniques, seed development represents the most suc-cessful and dominant form of plant propagation: sexual reproduction, which is typified by sperm-egg fertilization. Plants that do not have a sex life or that exercise other means are said to replicate asexually. The simplest but often unrecognized method of this type of reproduction is vegetative propagation: the development of a new plant from a growth or a detached portion of its parent. Other forms of asexual reproduction include fission, an action in which bacteria, unicellular fungi, and algae simply divide in two; budding, when a yeast cell sprouts daughter cells; and sporulation, when a reproductive cell, a spore, develops directly into a plant, as with algae and fungi.[16]

In simple terms, then, a spore is a lone reproductive cell capable of producing a mature plant without first combining with another cell. While this definition of "spore" is fairly straightforward, the term covers substantially different types of reproductive cells. For example, the spreading hyphae (threadlike growths) of bread mold that have not mated with each other will send up slender stalks, each topped with a tiny container of spores. Upon release, each spore germinates and grows into new bread mold organisms. Yet if two differing hyphae are compatible (designated + and –) and bond, the result is the creation of a thick-walled capsule, which in turn develops a tiny container filled with spores. Thus, some plants may produce spores either sexually or asexually, but in neither case does the spore itself combine sexually with another before germinating. Other fungi, such as morels and various mushrooms, have this sexual/asexual spore-generation character as well.

Mosses and ferns also go through a sporulation phase, which is additionally linked to a sexual stage wherein sperm and eggs are produced— a process called alternation of generations. Initially, spores dispersed by these plants grow into visible plant bodies, each bearing sperm- and egg-producing organs: the fern forming a small two-lobed leaf and the moss sprouting a leafy vertical spike. Upon fertilization of the egg, a new structure arises—on the moss, it is a stem topped by a spore-bearing capsule; and on the fern, it is the familiar frond, which carries its spores in packets on its undersurface. Because these spores develop within a stage of sexual reproduction, they differ from the asexual spores of the fungi and the algae.

Unfortunately, within the lexicon of science the term "spore," in combination with various prefixes, has other applications, which can be confusing. These uses ignore the meaning of spore just cited and usually refer to various cellular bodies in a particular developmental phase (gametophytic) of the reproductive cycle. For example, *mito*spore is the technical name of the asexual spore of bread mold, whereas the sexually produced spore of the bread mold is a *meio*spore, and the body from which these meiospores develop is the *zygo*spore. Egg fungi may carry either *oo*spores or *zoo*spores, whereas various other fungi develop *asco*spores, *aecio*spores, *basidio*spores, or *uredio*spores. In fact, all plants that reproduce sexually employ "spores" of some type—even those bearing seeds. The important point to keep in mind is that when "spore" is used alone, it invariably refers to a cell that develops directly into a conspicuous plant structure without uniting with other cells. In the case of bread mold, this structure is a finished organism, and with mosses and ferns it is just one of the two plant forms produced in their generational cycle.

Seeds develop as the result of the sexual recombination of genes from within one plant, called self-fertilization, or from the genes of two separate plants, called cross-fertilization or cross-pollenization. In the pollenization of flowering plants (the angiosperms), the male reproductive cells, *microspores,* develop into pollen grains that sit within structures on the flower called anthers. When these grains mature and separate from the anther, they are transported to the female sexual structure, called the stigma. This occurs usually through insect and animal contact or from dispersal by wind or water. Upon reaching the stigma, the pollen grain seeks out its female counterpart, the ovule, which has developed from a *megaspore,* and fertilizes it—thus initiating the development of a seed. Nonflowering plants (the gymnosperms), such as the cycads, the ginkgoes, and the conifers, have neither anthers nor stigmas but bear male and female cones, or strobili (catkin-like structures), which carry out fertilization in essentially the same manner.

When an ovule is fertilized, three developments take place. A rudimentary plant form, the embryo, is created that carries all the genetic information and potential for a new plant. A storage tissue that nourishes the embryo, the endosperm, develops. And finally, a hardened outer layer called the "coat" forms, protecting the embryo from water loss and outside disturbance. Together, these three components form the seed. In angiosperms, an ovary wall encloses the seed coat, whereas in gymnosperms the seed lies exposed on the surface of a cone scale or, as in the yews, the embryo is surrounded by a fleshy coat and is called an aril.

Spores do not have an outer coat nor do most carry a food reserve that could extend their life. These factors force a spore-bearing plant to produce millions of spores so as to assure successful propagation from the few that manage to survive. When plants evolved into seed producers, this enabled various species to spread over territories and into climatic regions formerly hostile to lower plant life. Because the seed carried a protective coat and its own food supply, it could weather long periods of dormancy and could begin growing without initial dependence on outside nutrients. Also, because many plants reproduced through cross-fertilization, which introduced variation into species, this enabled such vegetation to adapt to new environments and subsequently become the dominant forms of plant life.

Sweet Potato • Yam

In shopping the produce section of our supermarkets, we will likely find a sign that reads either "sweet potatoes" or "yams." Regardless of the name, the vegetable is the same: a yellow to rust-colored oblong root that is fairly high in sugars. But unlike the shopper looking for sweet potatoes, anyone seeking true yams will be sorely disappointed; other than a few ethnic specialty grocers, our grocery stores do not sell real yams. The use of "yam" for the sweet potato is, in fact, a gross misnomer. Indeed, with the exception that both are vegetables, sweet potatoes and yams have very little else in common.

Botanically, all flowering plants fall into one of two major classes: the monocots, which are mainly herbaceous and have embryos with one "seed leaf"; and the dicots, which are most often woody and whose embryos have two "seed leaves." Sweet potatoes are dicots and yams are monocots, and this fundamental characteristic alone makes them significantly different plants.

Sweet potatoes belong to the *Ipomoea* genus of the morning glory family, Convolvulaceae. Their species name, *batatas,* is an American aboriginal word from which "potato" was derived. While the sweet potato resembles a potato, it is not a tuber and does not develop "eyes." It is a true root that has swelled into a potato shape, sometimes round but more often oblong. Two types of sweet potatoes predominate: a somewhat dry and mealy variety used primarily for flour and livestock feed, called a Jersey, and the sweeter, softer variety grown for human consumption. The latter, rich in starch that breaks down into simple sugars, is the one we call a "yam." Normally a perennial plant, as a crop it is cultivated as an annual.

Sweet potatoes, no longer known to grow wild as they once did in South America, are cultivated extensively throughout the wetter temperate regions of the world. They are generally not a staple food in most diets, but on some Pacific islands their consumption is quite high. Outside the tropics, sweet potato crops have been grown as far north as Spain. Introduced to the United States in 1650, the sweet potato is raised principally in the South and the East. The edible sweet potato, or "yam," is harvested extensively in Louisiana and in a few states along the eastern coast. The Jersey variety is grown in the more northern states of Virginia, Delaware, Maryland, New Jersey, Kansas, and Iowa.

The true yam, genus *Dioscorea* (family Dioscoreaceae), is a tuberous plant like the potato, and, whereas it is usually an underground vegetable, some edible species grow as aerial tubers. The word "yam" comes from the African *nyami*. Black slaves who, when brought to North America, saw the resemblance of the sweet potato to their native plant, bestowed *nyami* upon the new vegetable; hence, our misnamed sweet potato. In the same vein, *nyami* has also been misapplied to many other roots of the tropics. Of the 150-plus species of yams found worldwide, only about 10 are grown for food.

Normally medium-sized vegetables like the sweet potato, some yam species grow up to 8 feet long and weigh over 125 pounds. Almost all are perennial climbers, with vines that easily reach heights of 25 feet. To cultivate the yam, usually the whole tuber is planted: however, if the crop is small or the plant yields only one yam, the top of the tuber, containing the eyes, is cut off and buried.

Most yams are raised in the tropics and in Asia, where they provide an important source of starch. Not as palatable as the sweet potato, they sometimes have a disagreeable, bitter taste that can be removed only by boiling. Baking, frying, and grinding into meal prepare the more edible species, such as *Dioscorea alata* and *D. sativa*. In the United States only four native species thrive, none of which are grown for our tables. The wild yams of North America range from Rhode Island to Minnesota and Ontario, south to Texas, and eastward to Florida. Only in the South, particularly in Louisiana, are they a cultivated crop. They are not raised as food for humans or animals; their value lies in the cortisone, a steroid, that they contain.

Tree ● Shrub ● Bush

According to the United States Forest Service, a tree is a mature woody plant at least 12 feet high, with a single trunk whose diameter is at least 3 inches at a point 4.5 feet above ground. Although this is a popular definition, other authorities consider that a tree need be only 8 feet tall and have a trunk diameter of 2 inches. Still others hold that 15 feet is necessary. And some say that a tree will usually have one or more perennial stems at least 3 inches in diameter at breast height at maturity and have a more or less definitely formed crown of foliage at least 16 feet above ground. But as a general consensus, most authorities consider 20 feet to be a reasonable minimum height (some stipulate a minimum diameter of 4 inches several feet above ground level). However, these requirements all assume good growing conditions. A tree that has the potential to reach any of the previous heights may not, in poor soil or under harsh conditions, such as on the Canadian tree line, grow more than a few feet tall. Stunted as it may be, it is a full-grown miniature of its more robust relatives to the south.

On first consideration, the condition that a tree be woody would seem reasonable. Oaks, pines, yews, willows, and so on, are all woody, but palm trees are not, nor are banana trees. Palm trees, such as the date, the coconut, and the royal, are not considered woody plants because they do not produce bark or annual rings or accumulate secondary xylem (wood-producing tissue). Neither do they have specialized structures in their trunks to transport water and nutrients—the entire trunks are involved in the process. Furthermore, almost all palms grow above ground from single points on their crowns (an exception is the branched Doum palm of Egypt and the Sudan). Yet they attain significant heights on single stems and are considerably more than three inches in diameter.

The banana tree is something else. The term "tree" is often applied as a popular convention because of the plant's single-stalk resemblance to other trees—principally, the palms. Despite this similarity, the banana plant is not a true tree, although under some classification schemes it is categorized as a "false-trunk" tree. It can grow over 20 feet tall; its single stalk is greater than 4 inches in diameter, but it has no trunk! Its herbaceous stalk is no more than leaves and leaf bases that have tightly wrapped themselves around each other to form a heavy shaft, which is terminated by a banana bunch. Much the same can be said of tree ferns. As true ferns, these tall (up to 60 feet), sometimes branched, palmlike "trees" have trunks that arise from rhizomes (horizontal underground stems) and do not increase in diameter as the plants grow.

The qualification that a tree be a plant with only a single trunk is also misleading. Birch trees, silver maples, and black willows often occur with multiple trunks arising at ground level. And consider the incredible quaking aspen in Utah that covers 106 acres. Called the second-most massive living organism in the world, this unique plant, with but a single root system, has developed an estimated 47,000 tree trunks. Is the whole organism one tree, or are there thousands of individual trees? Then there are the many single-stemmed shrubs. The blackhaw shrub normally has multiple stems and develops a crown close to the ground, yet it can be grown 15 feet high on a single trunk and is then spoken of as a tree. The same is true of the 20-foot high *Siebold viburnum*.

What is a shrub, then? Most often, it is said to be a low-growing woody plant having several stems, rather than a single trunk. That most are low growing is true, but there are some exceptions. Certain lilacs can grow taller than some trees, and woody vines like the wisteria, considered a shrub, can attain considerable heights. Although most references set 3 meters (about 10 feet) as the maximum height for shrubs, some raise this limit to 6 and even 10 meters.

Not all shrubs are entirely woody, either. Some are intermediate between woody and herbaceous. Often termed "semiwoody," "semi-herbaceous," or "subshrubs," they typically have crowns of herbaceous foliage and stems of at least partially woody tissue. The teaberry and the bittersweet nightshade are examples of such borderline woody shrubs.

What can be said, then, of the difference between trees and shrubs? The mountain laurel and the rhododendron, depending upon where they grow, could be called either trees or shrubs. Under different circumstances, certain species of trees and shrubs can have either one trunk or many. Trees need not be woody, nor shrubs entirely so. And there are shrubs that grow taller than some trees. As one naturalist put it: "It might be simpler to think of a shrub as any woody [?] plant that is not a tree."[17]

In truth, the distinction between trees and shrubs is not dependent on any one or several characteristics but is more a tool of convenience. Each term signifies the extent—more or less—to which a plant resembles the typical "tree" or a "shrub." Some trees are shrublike, and some shrubs are treelike. But regardless of which term we choose, both indicate that the plant is not herbaceous. Herbaceous plants have very little woody tissue in their stems, and all die down to the ground each year. Usually, they are termed annuals, biennials, or perennials. In the subshrubs, which have woody lower stems but herbaceous branches, the stems survive wintering, while the upper growth dies back.

The term "bush" is sometimes used interchangeably with "shrub." One author of a book on shrubbery says that shrubs are simply bushes. Unfortunately, he neglects to tell the reader what he means by "bush." Probably the best way to look at the term is if one considers its adjectival form, "bushy." A bush is a shrub or a small cluster of shrubs appearing as one plant that has a "bushy" look. That is, it is full and low-growing, has dense foliage, and usually branches out close to the ground. Thickets are made up, in part, of a dense growth of bushes. The taller shrubs with high crowns of foliage are not considered bushes, nor are woody vines, which are classified as climbing shrubs.

3

TERRESTRIAL INVERTEBRATES

Insect ● Spider ● Crustacean

It is commonly assumed that all crawling, flitting, burrowing, biting, irritating, and often repulsive little creatures are either insects or spiders, spiders being those that look spidery and spin webs, and all the rest insects. Not so.

Insects and spiders are members of the animal phylum Arthropoda (meaning joint-legged), with insects making up the class Insecta, and spiders, in part, comprising the class Arachnida. However, these are not the only invertebrates in the phylum. Arthropoda has six other classes, all with members commonly mistaken for insects.

Besides the insect and the spider classes, Arthropoda also includes Crustacea (lobsters, barnacles, shrimp, etc.); Chilopoda (centipedes); Diplopoda (millipedes); Pauropoda (pauropods); Symphyla (symphylans); and Onychophora (velvet worms—sometimes considered a separate phylum). It should be noted that although crustaceans are traditionally considered arthropods, many authorities feel that they deserve their own phylum. The pauropod and the symphylan are very small creatures, less than half an inch long, that resemble the centipede. Velvet worms are primitive caterpillarlike invertebrates found only in Asia, Central America, and the southern hemisphere. The more commonly encountered centipede and millipede are examined in a separate discussion. Two other terrestrial "noninsects," not of the Arthropoda phylum, are the land snail and the slug (essentially a snail without a shell). While seldom mistaken for insects, their true classification is sometimes forgotten. Snails and slugs are mollusks, invertebrates belonging to the class Gastropoda (meaning "stomach-foot").

Insects encompass a seemingly endless array of anatomical variations. Their sheer number (700,000 identified species) and wide assortment of forms would seem to virtually preclude any mutual attributes, yet all insects share three distinctive features. Because many insects change form as they develop, consideration is given only to adult characteristics. The three principal features are a body separated into three distinct parts (head, thorax, and abdomen); three pairs of legs, each with five major segments; and a tracheal system of respiration—similar to that of centipedes and millipedes. Characteristics frequently found in insects, but not universally, include two pairs of wings; a pair of antennae; compound eyes, plus two or three simple eyes; and development through larval and pupal stages.

In some classification systems the insect orders Collembola (springtails), Diplura (two-tailed bristletails), and Protura (proturans) are set apart from the others and are not regarded as true insects because their mouthparts are enclosed in a pouch. Two of these primitive, wingless arthropods, the springtails and the proturans, also lack eyes; and the proturans, lacking any antennae, use their enlarged, hairy forelegs as sensory organs.

Spiders are arthropods, sharing the arachnid class with mites, ticks, scorpions, and a few lesser-known members. One arachnid, the harvestman or daddy long-legs (a term the British reserve for the crane fly), is often mistaken for a spider but can be distinguished by the long spindly legs of most species, the segmented abdomen, one pair of eyes, and an absence of a waist. The spider has only two body parts: an abdomen and a cephalothorax (a conjoined head and thorax). This characteristic and its 8 legs with knees (a feature absent in insect legs) are its most obvious features. The spider also possesses 2, 4, 6, or 8 simple eyes (8 being the most common), poisonous fangs, silk glands, and pedipalps—short appendages near the mouth used for feeding and copulation. And unlike insects and crustaceans, spiders lack antennae and true jaws. Respiration is accomplished in one of two ways: in the more primitive spiders, it is performed by a system of stacked blood-filled plates called "book lungs," around which air circulates. The other method uses a simple trachea, which provides better air exchange for the more advanced and active families that have a higher metabolism. Spiders also differ from most crustaceans and other insects in developmental growth; they do not go through a larval stage but are born resembling their parents—as are we.

It may seem odd that crustaceans are included in this category, considering that almost all are aquatic, but two terrestrial forms, the pillbug and the sowbug, are so common and insectlike that most people mistake them for insects. The term "crustacean" refers to the flexible, hardened outer skeleton, called a carapace, that most possess. Compared to the other arthropods, some crustaceans grow to an immense size: eastern lobsters can reach 35 pounds and 3 feet in length. Crustaceans dwell almost exclusively in marine and freshwater environs, rarely making their home on land, whereas insects (except for the few aquatic forms) keep to dry land.

Physically, crustaceans have several features that make them unique. All have at least three pairs of appendages that act as jaws or feeding tools, two pairs of antennae, gills or pseudotrachae for respiration, and eyes that are compound and often stalked. Many also begin life as nauplius larvae, and those that do not are thought to have ancestors who did. A nauplius larva is distinguished by an unsegmented body, three pairs of appendages, and often a single eye. Some barnacles (also a form of crustacean) that are parasitic on marine worms also lack mouths and anuses.

Pillbugs and sowbugs, the two insectlike members of Crustacea, are those small, grayish, lozenge-shaped "bugs" commonly found on damp ground or hiding under rocks and logs. Sometimes called woodlice, each has seven pairs of legs extending from beneath a short row of plates that forms its oblong body. Pillbugs can be distinguished from sowbugs by their shorter and more rounded shape, their round tail ends, and the ability to roll themselves into a ball—a means of defense. Pillbugs also favor grassland and deserts, whereas sowbugs prefer humus and leaf litter.

These two terrestrial invertebrates have successfully adapted to life on land by modifying several features still found on their aquatic relatives. Although other crustaceans use gills, pillbugs and sowbugs use an organ for respiration that, while resembling tracheae, functions more like the spider's lung; their legs are better adapted for walking than for swimming; they reproduce by internal fertilization; their eggs are retained in a brood pouch until the young fully develop; and their mouths do not filter food but are designed for chewing.

The following listing summarizes the major differences between insects, spiders, and crustaceans.

Feature	Insect	Spider	Crustacean
Body parts	Three	Two	One to three
Pairs of legs	Three	Four	Varies
Eyes	Compound and simple	Simple	Compound
Antennae	One pair on most	None	Two pairs
Respiration	Tracheal	Book lung or tracheae	Gills or pseudo-tracheae

continued on next page

continued from previous page

Feature	Insect	Spider	Crustacean
Development	Direct, nymph, or larval	Direct, no stages	Usually larval
Poison device	Rear of abdomen, if present	Fangs	None
Wings	Most with two pairs	None	None
Habitat	Land, seldom aquatic	Land	Water, rarely terrestrial

Bee • Wasp • Hornet • Ant

Bees, wasps, hornets, and ants all belong to the insect order Hymenoptera, a name coined for the linked, membranous wings most species bear. Many also possess venomous stingers, appendages used for defense and stunning prey. Although this weaponry makes them less-than-welcome guests, many species are quite valuable. Wasps, and bees in particular, pollinate more fruits and vegetables than does any other group of insects. Another interesting facet of these tiny creatures is their communal behavior. Ants, and some bees and wasps, form colonies with castes and permanent divisions of labor—an exceptionally rare lifestyle among insects.

Ants, bees, and wasps comprise most of the suborder Apocrita, insects that are distinguished by a constriction near the base of their abdomens, and ovipositors (egg-laying apparatuses) that are modified to pierce or sting. The only other suborder of Hymenoptera is Symphyta, which contains the sawflies and the horntails. Apocrita is divided into 11 to 17 superfamilies (depending on the authority), one of which, Apoidea, harbors all the bees and another, Formicoidea, the ants. The other superfamilies contain various families of wasps and wasplike insects.

For most of us, distinguishing bees from wasps is seldom a problem. Bees, forming but a single superfamily of stout-bodied species, are much alike, making "bee" a fairly unambiguous term. "Wasp," however, is used in so many ways by various authorities that any narrow definition of the term would likely be open to question. Some authorities contend that "wasp" covers any stinging member of Hymenoptera that is not a bee or an ant; however, this would exclude the 1,000-plus cuckoo-wasp species that do not sting. Some make a distinction between wasps, horntails, and the ichneumon, whereas others claim that all three are wasps. Many contend that only three superfamilies (Scolioidea, Sphecoidea, and Vespoidea) contain "true" wasps and that the "wasps" of the other superfamilies are only nominal wasps or, at best, allowed exceptions. Other scientists, while recognizing that several superfamilies are needed to properly classify wasps, disagree on which features should govern inclusion. One authority, while adhering to the traditional view that wasps belong only to the suborder Apocrita, inexplicably refers to the horntails as "large wasps."[1] Clearly, wasps are not easily pinned down. What most authorities do commonly agree on, however, is that those of Scolioidea, Sphecoidea, and Vespoidea are unquestionably wasps. Here we find the yellow jackets, the hornets, and the mud daubers, plus many lesser-known forms.

The odd duck among all the superfamilies is Formicoidea, which harbors all the ants. One might imagine that ants would be ranked higher (perhaps as a suborder), considering that Formicoidea's 297 genera contain over 8,800 species spread throughout the world. A higher ranking might also appear appropriate, in that ants seem so obviously distinct from those of the other superfamilies of Hymenoptera. Yet, physically, ants are very wasplike but for four main differences. As with bees, the ant's antennae bend at the "elbows"; the waist, or the second, pinched abdominal segment, usually carries a hump; the mandibles are elongated for fighting and capturing prey; and, one of the most important features of all—found on most ants but not visible to the naked eye—is the metapleural gland located just before the waist on the rear of the first abdominal segment. This tiny organ, consisting of a pair of cell bundles, produces an antibiotic that is used to destroy bacteria and fungi and also to repel attacking enemy ants.

It may seem odd that ants are even included in this discussion, considering that their features and their habitats appear markedly different from bees and wasps—except for the rarely encountered winged queens and males, ants are seldom mistaken for wasps or bees. Yet several wingless wasps so closely resemble ants that more often than not, they are mistaken for members of Formicoidea. One, the female velvet ant, despite its name and its similarity to ants, is really a wasp—it lacks a hump, enlarged mandibles, and a metapleural gland.

As indicated, hornets belong to one of the three widely recognized wasp superfamilies: Vespoidea. Unlike the ant, which has attained standing apart from the wasps, hornets have not. In fact, they belong to the same family, Vespidae, as do the yellow jackets. To make the distinction between wasps and hornets is much like saying that dogs and cocker spaniels are different animals. Hornets are wasps—contrary to those who portray them as separate insects.

In Europe, the term "hornet" refers to essentially one species, *Vespa crabro*. This social wasp gained a widespread reputation because of its distinctive coloring (orange and browns) and large size. Bearing stout bodies, about 1⅛ inches in length, they are much larger than any of the other social wasps. The hornet, originally a European wasp, was introduced into the United States in the mid-nineteenth century, where its populations have remained confined to the eastern states. As the hornet was the largest wasp around—a trait that struck fear into most whom it met (actually, it is relatively passive and mild-mannered)—its name soon took on fearsome connotations. This arose mainly because people, long familiar with the smaller, aggressive native wasps, assumed that the hornet shared these wasps' ill-tempered nature. In turn, unwittingly or not, people began using "hornet" for many of their smaller wasps, linking them to the hornet's newly acquired and undeserved reputation. In main, these smaller native wasps are the social, paper-nest building species: typically, the yellow jackets such as *Vespula pensylvanica, V. maculifrons,* and *Dolichovespula maculata.*

D. maculata, also called the white-faced hornet because of its yellow-white markings, is squatter than those of *Vespula* and lacks the characteristic yellow band that girds the *Vespula* midsection. These squat yellow jackets also build aerial nests in trees, under eaves, or beneath bridges, whereas the yellow jackets of *Vespula* build their paper nests in or close to the ground. Although "yellow jacket" is closely linked to the species of *Vespula* and *Dolichovespula,* the term has become somewhat colloquial and is now often applied to any yellow and black wasp (a practice that further aggravates the confusion over "hornet" in the United States).

Bees are ancient descendants of wasps. Millions of years ago, they separated from the Sphecoids and evolved to the point where they now merit their own superfamily. Traits unique to various bee species include the secretion of wax to make hives, the storage of honey and pollen (a few vespid wasps do gather pollen and make honey), and a highly developed means of communication. Whereas these features suggest a considerable amount of communal cooperation, 90 percent of bee species are solitary. Yet those bees that do form colonies comprise the only known insect group able to sustain a succession of generations with stored food.

Physically, bees differ from wasps in several ways. The first sections of the hind tarsi (the last leg segment) of almost all bees are enlarged and somewhat flattened for pollen gathering. The abdomens are never stalked, or "wasp-waisted." The proboscises are more elongated for reaching plant nectar. And, unlike those of wasps, the leg or body hairs of bees are specially branched or fringed for picking up pollen. The one other difference, and a significant one, is that wasps feed their young animal matter, whereas bees provide their young with the same food they eat, pollen and nectar. Although a few wasps will continue to feed on animal matter their entire life, many switch to vegetation. Both bees and wasps are attracted to sweet liquids, such as ripe fruits and soda pop.

One feature of bees and wasps that is often cited as distinguishing the two is actually a long-standing fallacy. When they are contrasted, mention is frequently made of how a bee will lose its stinger in the flesh of its victim because the barbs on the stinger prevent it from being withdrawn. The wasp, it is then said, does not lose its stinger because it has no barbs, which enables the wasp to sting repeatedly. First of all, some wasps do carry barbed stingers, although the barbs are exceptionally small and the stingers can be easily withdrawn. The major error, however, is including all bees in the distinction. Native North American bees do not have robustly barbed stingers. Only the honey bee, which was introduced to the Americas by European settlers, has such a stinger and loses it upon attack—and often its poison gland and part of its intestine as well, leading to the bee's death within a few hours. The bumblebee and other North American bees, if so inclined, can sting repeatedly, just like the wasp; their stingers are no more barbed than are the stingers of hornets and yellow jackets.

Bug ● Beetle

Which of the following would you think is the only true bug: the Junebug, the ladybug, the mealybug, the red bug, the pillbug, the sowbug, or the waterscorpion? It is probably somewhat obvious that the correct answer is the one that appears least likely, the waterscorpion. Junebugs and ladybugs are beetles, the mealybug is a wingless scaled insect, the red bug is an arachnid, and both pillbugs and sowbugs are crustaceans.

Clearly, "bug" covers a number of different invertebrates. In fact, "bug," in the informal sense of the word, has become an accepted designation for almost all terrestrial invertebrates not readily identifiable. We frequently call millipedes, crickets, cockroaches, ticks, wood lice, and just about every other creepy crawler "bugs"—and with some justification. Considering the thousands upon thousands of different arthropods (the phylum under which all of the previous are classified) in the world, it is unlikely that one would be able to correctly identify more than a handful of different forms. Probable exceptions include the more familiar and easily recognizable bees, flies, spiders, scorpions, and such. The use of "bug," however, can be taken too far, as expressed in the erroneous observation that "Lobsters are bugs, a fact never mentioned on dinner menus, and rarely discussed by lobster lovers."[2]

Aside from the broad, and often indiscriminate, use of "bug," the term has a much narrower meaning, one that science uses to designate those insects assigned to the order Hemiptera. These insects, sometimes called "true bugs," are 40,000-species strong worldwide, with about 4,500 living in North America. "Hemiptera" refers to the dual construction of a bug's forewings and is one of the most telling characteristics used in its identification. The front wings of a bug are thickened at their bases and change to colored or transparent membranous structures toward the tips—the relative size of the membrane portions varying with each form. The rear wings, hidden under the front pair while the bug is at rest, lack the thickened leathery segment and are entirely membranous. When at rest, the wings fold over the back, with the membrane portion of the front wings overlapping. To fly, a bug beats both pairs of wings—a method of flight used by only a few beetle species.

A bug feeds on fruit juices, sap, fungi, and blood. Not equipped to bite, it pierces food with a hollow beak, which serves to suck up fluid. Those feeding on human blood make up a fair percentage of the bug population, and whereas disease can be transmitted through bug bites, the number of such infectious species is small and the problem largely limited to Third World countries. Yet citizens of more developed societies are not totally neglected by these Dracula-like insects. The bedbug, which, fortunately, is not a disease carrier, finds North Americans very appealing. In fact, some victims bitten repeatedly over many nights have been left weakened from excessive blood loss.

When a bug hatches, it closely resembles an adult, except that its wings and reproductive organs have yet to develop. This form of infant insect is called a nymph—or a naiad, if aquatic—and must go through successive molts before reaching adulthood. Grasshoppers, cockroaches, leafhoppers, and others develop the same way, a process called simple or incomplete metamorphosis. Beetles do not.

Beetles mature by going through a process called complete metamorphosis, the most common type of development among insects. They, along with butterflies, mosquitoes, bees, and ants, start out as eggs, then hatch into larvae usually called grubs, and undergo a pupal stage where they transform into adults. Unlike the bug, then, the newborn beetle in no way resembles its adult form (see "Nymph").

Beetles make up the largest order of insects in the world, Coleoptera. This immense order consists of over 300,000 known species, 10 percent of which reside in North America. As with "Hemiptera," "Coleoptera" describes its order's wings: sheathed. The front wings (elytra) on a beetle are completely thickened; some are soft and leathery, others hard and brittle. At rest, they generally cover all of the membranous hind wings and almost always meet in a straight line down the middle of the back (a notable exception, the timber beetle, has only stubby, token elytra). This feature is one of the more telling marks of a beetle. Because its elytra are virtually useless for flying, the beetle raises them up and out of the way of its flapping hind wings.

With rare exception, beetles do not suck up their meal through a tube, as do bugs. Instead, they attack their food using well-developed mandibles. Although several beetles possess large pincer-like mandibles that look extremely powerful and are capable of delivering a good bite, these insects are fairly harmless if handled gently—beetles do not particularly care for people as food. Most, in fact, prefer to dine on plants. They attack almost all parts: roots, stems, leaves, fruits, and seeds. A lesser number are scavengers, feeding on other invertebrates or carrion. A few others also exist as parasites.

Additional differences between bugs and beetles show up in their eyes, antennae, and legs. Adult bugs and beetles have compound eyes—the eyes from which the term "bug-eyed" is coined. These eyes each consist of a bundle of optical units crowned by multiple lenses. In addition to their compound eyes, most bugs have one to three much smaller eyes called ocelli, or simple eyes, which have only one lens each. Their purpose is not clearly understood, but they appear to distinguish light from dark and also may measure light intensity. A bit less reliable feature for telling the two apart is their antennae. Those of bugs have between 5 to 13 segments, whereas beetles' antennae are limited to either 10 or 11. Likewise, the tarsus, or last leg section, may or may not be a distinguishing appendage. A bug's tarsus usually consists of 2 or 3 segments, and a beetle's typically has 3 to 5 segments.

Centipede ● Millipede

Neither centipedes nor millipedes are insects. Insects come with six legs, no more, no less, whereas centipedes and millipedes, as their names would seem to suggest, should have either a hundred or a thousand legs (*centi:* "hundred," *milli:* "thousand," and *pede:* "feet"). Actually, the number of legs on both varies considerably, but it is many more than the six of insects. At one time centipedes and millipedes were grouped together in the class Myriapoda—denoting arthropods with many legs—which also included two similar "multileggers," the symphyla and the pauropod. Under current classification, each now resides in its own class, with centipedes making up Chilopoda and millipedes forming Diplopoda. These four multileggers, not at all closely related, are still collectively referred to as myriapods or, under some taxonomic schemes, as four classes in the superclass Myriapoda. Caterpillars and other insect forms with more than six "legs" do not qualify because only six of their many legs are true legs, and they are in a transitional stage, not in their final adult form.

Centipedes are the more familiar of the two, occasionally found inside homes seeking prey that feed on garbage and other organic material. Millipedes, which outnumber the centipedes almost 4 to 1 in number of species, are timorous creatures that prefer the soil and decaying plant litter, and if they are spotted, it is usually because the earth or covering material has been moved.

Other than their somewhat similar body shapes supported by numerous legs, centipedes and millipedes have very little else in common. Centipedes are quick-moving carnivores that may reach lengths of over a foot. The smaller species, with fewer but longer legs, are generally the fastest, some able to speed along at close to 2 feet a second. Their prey is commonly mites, snails, worms, silverfish, insect larvae, and other small insects. The familiar 2-inch-long light brown house centipede is harmless to humans, but a close relative in the tropics is reputed to have a bite as severe as a wasp sting. This centipede, and all the others, executes such assaults with its first pair of legs, which have been modified into curved, hollow, poisonous claws. Whereas many smaller species are essentially harmless to man, the larger centipedes of the tropics that catch and paralyze frogs, lizards, and mice with their venom can inflict extremely painful and dangerous bites on humans.

Centipedes and millipedes spend most of their lives out of sight of predators. Centipedes normally rest under cover during the day and emerge at night to forage for food. If suddenly exposed or illuminated, they quickly scamper away. Millipedes live much of their time safely burrowed underground or beneath dead vegetation. Normally not quick enough to evade danger by running for refuge, many rely on their hard shells and, for a few, the ability to roll into a ball for protection.

Unlike the predatory centipede, millipedes have little use for well-developed claws. Their principal tools of obtaining food, which occasionally may be dead animal matter, are robust jaws and, in some species, piercing and sucking mouthparts. Along with their protective shells, most millipedes also employ chemicals to repel or kill attacking predators. Along the sides of the millipede's body run two rows of stink glands that discharge a toxic fluid containing hydrocyanic acid: acid capable of peeling human skin. Large tropical millipedes, able to spray this solution up to 3 feet, have blinded not only hungry birds but children as well.

Figures vary concerning the number of species in each class. Chilopoda is estimated to have about 2,500 named members, whereas Diplopoda estimates run close to 10,000 species. With this many species, it is not surprising that the number of legs they bear also varies greatly. Centipedes have a minimum of 15 pairs and a maximum of 177. For millipedes, the number of pairs of legs ranges from 12 to 375.

Walking with this many legs has produced two forms of locomotion. Centipedes, with one pair of legs to a segment, wriggle as they move. Legs on opposite sides are alternately picked up and set down in groups of pairs, which produces an undulating motion. The short-legged millipede, carrying two pairs of legs to a segment, advances smoothly over the ground, with very little irregular body movement. The legs on both sides are in phase when shifted and cause a wavelike action that, at any one moment, can involve over two dozen pairs.

These differences, plus several others, are summarized in the following list.

Centipede	Millipede
Flatter body	Rounder body
Long legs	Short legs
Long antennae	Shorter antennae
Poison claws	No claws
One pair of legs to a segment	Two pairs of legs on each segment
Tracheae are branched	Simple tracheae are unbranched
Three pairs of jaws	One pair of jaws
No body venom	Ejects acidic body venom
Walks with a wiggle	Walks with a smooth glide
Moves quickly	Moves quite slowly
Carnivorous	Principally herbivorous
Genitals in rear	Genitals up front
Direct sperm transfer	Indirect sperm transfer

Dragonfly ● Damselfly

Fossil records show that 300 million years ago, during the Carboniferous period, long before the dinosaurs, dragonflies with wingspans of up to 2½ feet (the largest recorded insect ever) were hunting prey along the water's edge. Today, their largest relative, a damselfly with a wingspan of 7½ inches, still hovers above the water, watching for a meal, its basic body shape little changed over millions of years.

Dragonflies are known by several popular names: darner, sewing-needle, devil's darning needle, snake doctor, and horse-stinger. Despite the malevolent connotations of these nicknames and the myths attached to them, dragonflies can neither sew lips and eyelids shut nor sting. The larger species, whose family is informally called "darners," can, however, deliver a fair but harmless bite. In fact, Odonata, the order of dragonflies and damselflies, refers to their biting mouthparts: toothed mandibles, which occur on most adults.

Odonata is an order of about 5,000 species, sometimes divided into three suborders but most often just two: Anisoptera, the dragonflies, which have "unequal wings"; and Zygoptera, the damselflies, with "paired wings." The distinction is quite noticeable. Whereas the hind wings of the dragonfly are broader than its front wings, both wings of the damselfly are equal in size and, unlike dragonfly wings, are usually stalked or narrowed at the base. Also evident is the position in which the wings are held at rest; the dragonfly holds its wings horizontally out to the side, whereas almost all damselflies keep their wings vertical, back-to-back, and pointed to the rear, close to the body. In flight, both can move each of their four wings independently, facilitating either forward or backward movement.

The larger wing size of the dragonfly, coupled with a stout and powerfully built body, enables this predator to feed over a far larger area than the more slender and delicately built damselfly. Some large (5-inch long), far-ranging "dragons" have been found over deserts and at sea, in some instances 200 miles from shore, sitting on ship railings. Damselflies, being weaker fliers, stick relatively close to home, their flights regularly interrupted by rest stops on vegetation. Frequently alighting and being hindered by slow flight make them much easier prey than the quicker dragonfly. As for speed, it is not uncommon for the more elusive dragonfly to exceed 10 mph, and an even faster species is thought to approach an incredible 35 mph.

The eyes of Odonata are huge. On some species, they comprise over half the bulk of the head. On the dragonfly, these eyes either meet or are never more than an eye-diameter apart and often follow the contour of the head. In contrast, damselfly eyes are always widely separated and usually look like large beads on the side of the head. Sitting on an easily swiveled head and endowed with acute vision, these large compound eyes have enabled both insects to become very adept hunters—often effortlessly capturing prey in midair with their legs. They are not at all picky eaters; their quarry ranges from the almost-invisible midge to small frogs plucked from the ground.

The aquatic larvae of these two predators, properly called naiads, are also aggressive hunters. The naiad carries a jointed structure in front of its face of modified mouthparts armed with hooks, and it can very quickly extend this weapon forward to capture prey. This apparatus, sometimes up to one-third the body length, flashes out in microseconds, snaring worms, tadpoles, crustaceans, aquatic insects, and small fish. When a naiad transforms into an adult, this appendage is lost. To seek out prey, the dragonfly naiad propels itself by drawing water into its rectum and then forcefully expelling it like a jet. Curiously, this anal chamber also holds its respiratory gills. A damselfly larva is quite different in both respects. Its leaflike gills are located outside the body at the tip of the abdomen and, in concert with the body, undulate from side to side, moving the naiad forward in a swimming movement. Depending on the species, food supply, and temperature, life in this aquatic stage may last from 30 days to 5 or 6 years. During this time a larva may molt from 9 to 15 times before emerging as an adult.

Grasshopper ● Locust ● Cricket ● Katydid

Over the years, mankind has regarded the grasshopper and its kin as both an asset and a curse. Primitive cultures valued these insects as food, medicine, and even entertainment. Crickets in ancient China, highly prized for their song, were frequently kept in cages, where their chirping provided comfort at night. Yet these creatures have also wreaked devastation upon crops and forage throughout the world, often creating widespread famine in underdeveloped countries.

All are members of the Othoptera (meaning "straight-winged") order of insects, whose shared features include enlarged hind legs used for jumping, generally narrow and thickened forewings, and usually broad and membranous hind wings (a few species lack wings). With well-developed mouthparts for chewing, most are vegetarians. By rubbing various body parts together, most adults, notably the males, produce mating songs that are heard through auditory organs on their bodies or front legs.

Othoptera has two suborders: Ensifera and Caelifera, each containing insects popularly called grasshoppers. All members of Ensifera (except for the mole crickets) have very long, hairlike antennae, which in some species can be up to 4½ times their body length, whereas the antennae of Caelifera are typically quite short and thicker. Other distinguishing features concern their egg-laying apparatuses (the ovipositors) and hearing organs. Because of antennae length, these two forms are sometimes referred to as "longhorned" (family, Tettigoniidae) and "shorthorned" grasshoppers (most commonly, the family Acrididae). Coneheaded grasshoppers, meadow grasshoppers, and shield-backed grasshoppers are all "longhorns." But, interestingly enough, entomologists do not consider them or any member of Ensifera to be true grasshoppers. They contend that this use of "grasshopper" is a simple misnomer that arose because of their resemblance to true grasshoppers, the "shorthorns" of Caelifera.

Of the 5,000 species of grasshoppers in the world, 9 in particular are truly remarkable. These are the legendary migrating locusts, which form incredibly massive swarms—enough to darken the sky. In 1955 a swarm in Morocco measured 155 miles long and 12 miles wide, and in the United States during the late summer of 1875, a swarm in Nebraska is estimated to have contained 12.5 trillion members, which covered close to 200,000 square miles. Not only do they develop huge swarms, but all are capable of traveling tremendous distances. In 1952 one horde flew 3,000 miles from Africa to India. Often at the mercy of the wind, they are carried aloft during the day and then settle at night to feed. Such plagues may last only a few months or several years; the last great locust swarming began in 1950 and did not end until 1962. Swarms have been estimated to consume the same amount of food in one day as 1.5 million humans, and wherever they settle, the land is stripped, with only barren branches and twigs left behind.

But this gregarious swarming behavior is not the only unique characteristic of the locust; strangely enough, a significant physical alteration takes place before swarming occurs. Normally, the locust is an unexceptional solitary grasshopper, living as such for many generations, just like its nonswarming brethren. But if overcrowding begins and individuals start bumping into each other, and if the temperature, the humidity, and the food supply are appropriate, hormonal changes occur that physically transform these species. Originally green to sandy colored, the locust changes to black, with a conspicuous pattern of yellow or orange markings. The pronotum, a shield over the thorax just rear of the head, changes shape slightly, and the wings lengthen. These developments do not happen at once but over a couple of generations, during which time individuals begin amassing, as their populations dramatically increase. However, if conditions should change, or when the locusts complete their migration, they will revert to their solitary phase by again changing physically and behaviorally.

Reference is sometimes made to the infamous 13-year or 17-year locusts that appear close to these intervals in noisy summer swarms, often settling in trees. Despite their name, they are not locusts or even a close relative. These are cicadas, members of the same insect order as bugs. Often erroneously called locusts because their swarming masses number in the tens of thousands, they all tend to emerge from their underground habitats beneath trees at the same time. These buzzing and clicking large-beaked insects resemble huge flies more than they do grasshoppers. But like the locust, their appearance can mean considerable damage to plant life, particularly to those fruit trees in which females deposit their eggs.

Crickets are all members of the suborder Ensifera, despite the few misnamed species in Caelifera, such as the pygmy sand cricket and the pygmy mole cricket. In distinguishing crickets from the true grasshoppers, we should also apply the following distinctions to the other members of each suborder.

As pointed out, true grasshoppers have short antennae—less than thirty segments. Antennae of crickets are longer, all having more than thirty segments. The ovipositor of the cricket, like those of most members of Ensifera, is long and sword shaped, whereas that of the grasshopper is short and inconspicuous. The tympanal, or auditory, organ on a cricket is located on the fore tibia, unlike that of the grasshopper, which is borne on the first abdominal segment. A cricket produces sound by rubbing its right and left forewings together. A grasshopper does so by rubbing its hind femur against the forewing. Also, a cricket generally has a shorter body in relation to its hind legs than does a grasshopper. And the winged cricket, but not necessarily the other members of Ensifera, lays its wings across its back, whereas a grasshopper rests its alongside the body.

"Katydid," as the word is used in North America, refers to several, usually green, often broad-winged, tree-dwelling Ensifera insects of the Tettigoniidae family that are often called bush-crickets elsewhere. Armed with exceptionally powerful hind legs, many do not bother to use their wings to fly but simply leap from place to place. "Katydid" was coined for the loud call that males make at dusk, sounding at times like "katy did she did," and the term is sometimes used in place of "grasshopper" in the names of the longhorned "grasshoppers" previously mentioned. It is also sometimes improperly applied to crickets, as well as a few more-primitive species of the suborder. Some British authorities, besides using "katydid" for those species singled out by North Americans, include all members of Tettigoniidae, even though not all produce the typical katydid call. This widespread family includes species found not only in North America but in Europe, Africa, and various tropical areas as well.

Mite ● Chigger ● Tick ● Louse

This is quite a diverse group of arthropods. Composed of insects and arachnids, it contains members ranging in size from the microscopic 0.004-inch follicle mite to the relatively gargantuan 1.2-inch-long tick. All four forms include species that live as blood-sucking parasites that transmit diseases, cause infection, or induce irritation. Other species, a small minority, are beneficial to their hosts as scavengers, cleaning away debris and unwanted secretions. Curiously, most have a preference for only one or two types of hosts; the scaly leg mite that attacks the feet of birds is not found on humans, nor does the hog louse feed on sheep—most are fairly host-specific. But not all feed on flesh. Many favor processed foods or wild vegetation, whereas others infest citrus and field crops.

Mites, chiggers, and ticks are arachnids (of the order Acarina or subclass Acari—depending on the classification scheme employed), which, unlike spiders of the same class, do not have divided bodies. As arachnids, they each lack antennae and have eight legs—although compared to spiders, their "larvae" start out with only six. Mites, diverse and colorful creatures, constitute all but three (those of ticks) suborders of Acarina. Depending on the species, they either feed as parasites on plants or animals (both vertebrates and invertebrates) or exist as free-living dwellers in the soil or water.

Those living on vegetation, like the spider mite, the brown mite, and the gall mite, suck up sap and lay eggs in the tissue of a wide variety of plants, creating a host of problems—many very serious. Fruit may discolor or drop prematurely, foliage can turn yellow and fall off, or buds may fail to develop.

Better known, and probably more detested, are the parasitic and infectious mites that favor animals, particularly humans. The human scabies mite, only 0.01 inch long, burrows under the skin and secrets a toxin into the flesh, causing a rash and itching. Sometimes called "barber's itch," this infection can fester for a month before clearing up. Then there is the follicle mite that lives on the human face in oil glands and hair follicles, particularly around the eyelids and the nose—not a pleasant thought. Other mites prefer poultry and cattle, such as the mange mite that infects domestic livestock with the disease mange.

One group of mites that has especially incurred contempt is the chiggers, also called jiggers, red bugs, or harvest mites. These are six-legged larvae from about 20 species of the mite family Trombiculidae. The chigger attacks a variety of animals, not by burrowing into the skin but by attaching itself to a pore or the base of a hair. Swelling and itching usually follow, with the possibility of fever within 24 hours. The flea known as a chigoe is often confused with and called a chigger because its name sounds similar. This subtropical insect usually burrows under the skin of the feet in humans and animals, producing sores.

Ticks, much larger and less colorful relatives of mites, feed only on land vertebrates as blood-sucking parasites. They exist in two distinct body forms for which they are named: the soft tick that feeds on chicken blood, sometimes needing only a couple of minutes to dine before dropping off; and the hard tick that attacks mammals and may hang on for days or weeks, gorging itself with blood until its body becomes dramatically distended. Although both forms carry disease, the hard tick is particularly dangerous. Children have become permanently paralyzed and consequently died of tick paralysis, the result of a tick tightly attaching itself to the base of the child's head and injecting nerve toxin. Other ailments inflicted by ticks include itching, swelling, anaphylaxis, and viral diseases. Rocky Mountain spotted fever, spread by the wood tick and the dog tick, is a type of rickettsia disease that, while treatable, has resulted in a mortality rate of 3.2 percent. Ticks also transmit Q fever and the lately prevalent Lyme disease.

The tick anchors itself with a grooved, dart-shaped, blood-sucking device located just under its mouth. Its backward-curving "teeth," aided by a gluelike secretion, grip so tightly that if the tick is forcibly removed, its head or its mouthparts will often remain imbedded in the skin. Mites do not have such an apparatus but rely on various other adaptations of mouthparts that suit their feeding requirements. Mite bodies also have adapted to these needs: the follicle mite is pen shaped for easily burrowing into pores; the scabies mite is round, with its short legs faced forward, for burrowing under skin; and the chigger mite has long legs for quickly grasping onto passing hikers. Ticks, however, are all fairly uniform in shape, looking like tiny leather sacks with legs.

The louse is a wingless insect (six legs, antennae, three body parts). All species are external obligate parasites that feed on mammals and birds but rarely produce any more distress than intense itching. The greatest harm they inflict is by transmitting diseases such as typhus, trench fever, and relapsing fever. Of all the lice, the best known and most detested are the three that infect humans. These are the body louse, often called a cootie; the head louse, whose eggs, called nits, inspired the label "nit-picker"; and the pubic louse, or crab, named for its resemblance to the crustacean. Both the head and the pubic louse attach themselves to their hosts by holding onto hair with their claws. The body louse prefers to fasten onto underclothing, frequently feeding from such a position.

Lice feed themselves in one of two ways: some, like the human louse, feed by sucking blood with piercing mouthparts; the others are chewers that use biting mouthparts to feed on skin, feathers, secretions, and blood. The sucking louse can be distinguished from the chewing variety by its broad and flattened body, a head smaller than its thorax (the body segment behind the head), longer antennae, and a single large claw on each leg. Whereas sucking lice restrict their hosts to mammals, chewing lice also infest birds. Lice, like mites and ticks, are also particular about where they dine. Most are very host-specific, living on only one or two species of animal. And as an endnote, plant lice are not true lice but aphids.

Butterfly ● Moth ● Skipper

Of all the insects in the world, none are more appreciated for their beauty than these three. The brilliantly colored and multipatterned wings of butterflies and skippers have not only inspired works of art but have themselves been turned into decorative pieces. In contrast, moths are generally less colorful and desirable, although many brightly colored, even iridescent, species do exist. Together, the butterflies, the skippers, and the moths make up the insect order Lepidoptera, which means "scale wings," a reference to the variously shaped, overlapping colored scales of cuticle that cover their wings.

Although skippers have a long ancestry of their own, in general discussions of Lepidopterans the skipper's close kinship to butterflies often prompts scientists and science writers to unite the two under the label "butterfly." Some authorities have assigned this grouping of butterflies and skippers the rank of order or suborder, calling it Rhopalocera, leaving moths to their own suborder, Heterocera. However, recent reexaminations of the order, particularly as to genitalia and wing venation, have led some scientists to devise a more "realistic" regrouping of the three into four suborders (not all authorities agree on this point, some even contending that there is no scientific difference between butterflies and moths). Under this reordering, the butterfly and the skipper superfamilies, plus the advanced moths, make up one suborder, with the more primitive moths forming the remaining three suborders.

Compared to the butterfly, a skipper has a thicker and stouter body (there are a few exceptions) and wings that are proportionately smaller. Also, skipper antennae sit farther apart on the head, and whereas they develop a club, like those of the butterfly, the antennae extend beyond the club in a slender segment that typically hooks backward. Not as noticeable are the two small spurs on the tibia of each skipper hind leg; the butterfy, in contrast, has only one spur to a leg. The larva (caterpillar) of a skipper, unlike those of the butterfly, is dull colored, smooth, and without spines and bristles. It is also easily identified by its large head, which is attached to a short, necklike constriction of the thorax (the giant skippers excepted).

At rest, butterflies hold both pairs of wings upright, whereas many skipper species will hold their front and rear wings at differing angles. But even more noticeable is the difference in their flight. Typical butterfly flight, although somewhat erratic in direction, is fairly fast, with a fluttering motion. Skippers tend to be swift fliers that do little fluttering but accent their flight with darting starts and stops, as if skipping—hence the name.

Unfortunately, unlike the many differences that distinguish butterflies from skippers, no one characteristic can be said to separate the two from the moths—every difference has an exception. (As a convenience, the use of "butterfly" in the remainder of the discussion also includes the skippers unless otherwise noted.) Yet the impression persists that such distinguishing physical characteristics do exist. One, color, has already been noted; and it must be added that just as some moths are quite colorful, conversely, there are somber-hued butterflies. But the feature probably most frequently pointed out as distinguishing the two is their antennae. All butterflies have some type of clublike enlargement at or near the antennae tips, whereas almost all moths have feathered or threadlike antennae that taper at the tips. The few excepted moth species, which display unmistakable butterfly antennae, all live in tropical America and the Australia-South Pacific area. Yet because the difference in butterfly and moth antennae is nearly universal, authorities have retained it as a benchmark for dividing Lepidoptera into Rhopalocera (knobbed or hooked antennae) and Heterocera (variable antennae). Obviously, this categorization is not foolproof.

To the entomologist or the knowledgeable butterfly collector, one almost infallible sign of a butterfly is the manner in which the front and the rear wings interlock during flight. With few exceptions, a butterfly secures its wings together by a lobe on the rear wing that locks into the underside of the front wing. Most moths, and a few species of butterflies from Australia, use a different method: each rear wing has a connecting apparatus with one or more bristles (called a frenulum) that engages a flap (in males) or short bristles (in females) on the front wing. A moth that does not have such a mechanism locks its wings together with a flap called a jugum. Those moths with clubbed antennae can be distinguished from butterflies by the presence of frenulums.

When the time nears for caterpillars to enter their pupal stage, most construct some form of enclosure. Moths are known for the silken cocoons they spin, and butterflies (although not skippers) for the hard-shelled cuticles that encase them—the result of their last molt. However, many moths, particularly those that pupate underground, never form cocoons, but instead excavate cells in the soil to house their transformation. With the exception of one family, skippers construct loose cocoons of silk and leaves that hang from branches or foliage. Butterflies, however, almost never spin cocoons, and those that do so construct pupa shrouds of loosely spun silk that only somewhat resemble cocoons.

The last distinguishing feature, which is often noted but only somewhat reliable, is general body shape. Aside from the skippers, most butterflies are slender and fairly smooth appearing, with only a handful bearing the fatter and often furry bodies common among moths. Besides physical variations, butterflies and moths also display several behavioral differences.

As noted, most butterflies hold both sets of wings upright, either together or at varying angles. In contrast, most moths spread their wings flat out to the sides or over their backs like a roof. Of course, exceptions to this crop up in both butterflies and moths. Finally, there is a bit of difference in their flight. The butterfly, as pointed out, flutters in a somewhat erratic pattern or in a start-and-stop skipping maneuver. Moths, more graceful fliers, generally make long, sweeping flights. Also, almost all butterflies prefer to fly during the day. Very seldom will any be seen after dusk and rarely at all into the night. Even overcast days will ground most butterflies. Moths are the dusk-to-dawn fliers, preferring the dark, yet, curiously enough, are attracted to lights. Exceptions to such behavior are the more brightly colored moths that prefer to fly during the day.

Caterpillar ● Grub ● Larva
● Maggot ● Nymph

From birth, insects follow one of three developmental paths to adulthood. The simplest, called ametabolous metamorphosis, is growth without any change in body form or function—which, as the word "ametabolous" implies, is really no metamorphosis at all. The common silverfish develops in this manner, the very youngest only a miniature version of the adult. Such ametabolic insects, all wingless, account for only a handful of insect species. All other insects go through one or more distinct changes in anatomy, like the amphibian that develops from an egg into a tadpole and then a frog. This transformation is true metamorphosis, a change involving at least two distinct body forms. In insects, this occurs in one of two ways.

The simplest of these is incomplete metamorphosis, technically termed hemimetabolous development. Grasshoppers, bugs, and dragonflies develop in this fashion. After emerging from its egg, the insect resembles an adult, except that its wings and often its genitals or other organs have yet to develop. Through successive molts, it gradually attains its adult form, sometimes losing organs or appendages along the way, as in dragonfly development (see "Dragonfly"). Insects in their immature preadult stage are called nymphs. Except for a lack of wings, terrestrial nymphs look much like the mature adults. Aquatic nymphs, called naiads, do not, nor, like land nymphs, do they breathe air. The naiad uses gills borne either internally or externally, which extract oxygen from the water. Nymphs of either form are sometimes referred to as larvae, a term in entomology (insect study) ordinarily reserved for a growth stage of the third type of insect development—holometabolic.

Holometabolous development involves complete metamorphosis. This, the most common form of growth, is marked by very significant changes. Beginning from an egg, the emerging larval form feeds almost constantly, continuing to grow until full body and functional development are achieved—the sole purpose of the larval stage. At this point feeding stops, and the insect enters a resting state called the pupal stage. Here, usually enrobed in a protective chamber it has made for itself and controlled by hormonal change, it undergoes a radical transformation in body structure. The result is an adult form far different from the larval. This is the egg-caterpillar-chrysalis-butterfly sequence of growth, which is also characteristic of ants, beetles, and flies.

Among those insects undergoing complete metamorphosis, only a few, such as members of Lepidoptera (butterflies and moths), Coleoptera (beetles), and Diptera (flies), have specific names for their larvae. The caterpillars of Lepidoptera are by far the most attractive larvae—most people would say the only ones—some sporting varicolored coats of dense soft hair with tufts of bristles running down the backs. Compared to other larvae, many grow quite large, producing sizable butterflies and moths. The most distinguishing feature of the caterpillar, however, is its legs. All species carry three short-clawed pairs up front, the same number as in their adult butterfly forms. In addition, almost all have between two and five pairs of false legs called prolegs, each bearing a ringlet of hooks called crochets, used for grasping leaves and stems. This device, serving as an aid to walking, is unique to the caterpillar. Those caterpillars lacking prolegs hold onto vegetation with their sticky undersides of suckers.

The larvae of beetles are commonly fat, pale, and sluggish, some with a sparse coat of hair, others quite naked. Most are called grubs—none would be called attractive. "Grub" is sometimes also applied to a few species of flies and, conditionally, to some bee species. Certain beetle larvae, because of their appearance or feeding habits, go by the names "worms" and "borers," such as the sugar-beet wireworm and raspberry cane borer. Likewise, "worm" and "borer" are also used for those caterpillar larvae with similar eating habits.

Whereas grubs usually develop thoracic legs, these legs, unlike those of the caterpillar, are not necessarily stubby; some, in fact, are distinctly jointed. Furthermore, unlike caterpillar legs that attach to an undivided thorax, grub legs emerge from a three-segmented thorax, a pair to each segment. But the most obvious difference between the caterpillar and the grub is the lack of prolegs or suckers on the grub—all locomotion is provided by the six legs up front, assisted by a wormlike movement from the rest of the body.

Fly larvae are commonly referred to as maggots, a term really appropriate for a particular type of larvae produced only by certain species. Other fly larvae are called bots, warbles, or grubs, or they have no particular name at all. But regardless of the larval type, one characteristic they all share is a lack of true legs, although many use pseudopods (prolegs) much like those of the caterpillar. Scientists call those larvae having blunt posteriors and pointed fronts "maggots" and will sometimes use "grubs" for the more barrel-shaped and fleshy types. Bots and warbles are fly larvae that live as internal parasites. The balance of fly larvae, such as that of the drone fly, which bears a long tail, or of the spined coastal fly, lack any special label—these are simply called larvae.

The term "maggot" is well known because it is the larval form of probably the most familiar of all fly species, the ubiquitous housefly.

Besides its identifying body shape, the maggot has a greatly reduced head that lacks a hardened covering. It also carries sharp mouth hooks, used for burrowing into plant or animal tissue. In order to breathe while tunneling through tissue, most have breathing pores located not only up front on their thoraxes, but also in their rears, allowing the burrowing larvae to draw in air from the tunnel's entrance to the rear.

It must be mentioned that although larva has been depicted as a particular insect form, the term is also used for the transitional stage of other organisms. Any animal that undergoes a metamorphosis from an immature form to an adult state can be considered larval. Crustaceans and sponges go through a larval stage, as do mollusks, sea urchins, and some roundworms.

4

AQUATIC LIFE

Clam ● Oyster ● Mussel ● Scallop ● Bivalve ● Lampshell

With the exception of the lampshell, these shelled creatures all belong to the phylum Mollusca. The lampshell is a unique form of sea animal, which by itself makes up the phylum Brachiopoda. A mollusk can be distinguished by the two distinct components that make up its body: a muscular head-foot structure used for feeding and locomotion and a visceral mass containing its reproductive, digestive, and circulatory organs. Most have fleshy coats called mantles, which surround most of their bodies and which, in the majority of species, also produce shells. Of Mollusca's seven classes, three are particularly notable: Gastropoda (snails, slugs, etc.), Cephalopoda (octopus, squid, etc.), and Bivalvia (clams, oysters, etc.).

Bivalvia is so named because its members all live within two valves, or shells. Often called "shellfish," they share this designation with other mollusks equipped with only a single shell, such as the abalone and the limpet. The shells, made from secreted calcareous crystals, provide protection from predators, allowing the animal to sit in open currents where flowing water supplies it with oxygen and nutrients.

Lampshells, so called because of their resemblance to ancient Roman oil lamps, are small, two-shelled marine animals that only superficially resemble mollusk bivalves. Up until the middle of the nineteenth century, their clamlike appearance had convinced science that they belonged with the other shellfish in Mollusca. On closer examination, however, their true nature was finally recognized, and it became necessary to create a new phylum for them, Brachiopoda, which has two classes containing about three hundred living species. External appearances aside, within the brachiopod's two shells is an animal quite unlike the bivalve mollusk. It has no two-part body with a head-foot arrangement attached to a large organ mass but rather a much smaller body with fringed tentacles, which it uses for feeding and respiration—mollusk respiration is accomplished with gills. One of the most noted features of the lampshell is the relationship of its body posture to its two shells. Its two shells lie above and below its body's symmetry, whereas in mollusks such as clams, the two shells lie to the right and the left of the body's symmetry.

Although clams are a specific type of mollusk, the term "clam" is often used as a catchall word and is freely bestowed on a variety of bivalves. In the United States it is applied to a host of edible shellfish—mussels, scallops, and oysters excepted. Conversely, in Scotland "clam" is usually limited to scallops, whereas in other countries only mussels qualify as clams. Because of the indiscriminate use of "clam," clams may appear to exhibit a variety of contrasting features, but even a cursory look at a true clam easily dispels such a notion.

The true clam has two adductor muscles (the edible part of the clam that closes the shell) and two siphons. The siphons are fleshy, tubular appendages that poke out of the shell, one taking in oxygenated and nutrient-bearing water, the other expelling waste. These appendages allow the clam to live comfortably hidden under the sand. In some species (e.g., the soft-shelled clams), the two siphons are completely enclosed by a single muscular membrane; in others (e.g., the razor clam), the two are quite distinct. A few clams even have "eyes," or at least light-sensitive organs on the ends of their siphons. In larger species, such as the giant clam of the Indo-Pacific waters, the siphons are greatly reduced. These large clams live above the seabed, often sitting wedged in coral or held in place with thin, exuded byssal threads, strong, silky filaments. To move about, the clam uses a muscular "foot" that protrudes from between its valves—Bivalvia's alternative name, Pelecypoda, meaning "hatchet foot," derives from this feature. Although all clams share these features, their anatomy is so varied that about a dozen superfamilies are required to properly classify them.

Oysters are marine bivalves belonging to the superfamily Ostreoidea and are set apart from most other bivalves by their lack of siphons. The oyster acquires food and oxygen, and expels waste, by drawing water across the opening of its two shells. This group of shellfish, which includes the edible oyster—never considered a clam—differs markedly from mussels and scallops. Oysters generally have an irregular shape that ranges from oblong to somewhat round, and their rough shells are sometimes encrusted with barnacles. The two shells of the oyster, one flat and the other convex, close by means of a single muscle located midway between its two ends. Early in life, the oyster attaches itself to rock by standing on end and secreting a sticky material onto the surface. It then lowers itself into this cement, where it will remain anchored the rest of its life, doing away with any need for a "foot."

Oysters that produce the valued pearl are not edible shellfish, nor are they true oysters. Although called a "pearl oyster," this bivalve, which makes up the family Pteriidae, is much closer to the mussel in character, attaching itself to rock by using excreted threads rather than cement. The pearl oyster's outline is also more symmetrical and its shell surface somewhat scalloped and marked by concentric ridges. One other group of so-called oysters is Malleidae, a family noted for its extended hinge line.

Mussels occur in two very similar forms: marine (order Mytiloida) and freshwater (order Unionoida). Their chief difference lies in the construction of their gills. Common to both, however, is the manner in which they explore their surroundings. The mussel crawls about using its "foot," searching for a suitable place to rest and feed. Finding such a site, it excretes strong, thin, sticky filaments onto the rock, where it holds fast until it decides to move on. To change location, the mussel may break the threads and resume crawling by foot, or it may exude new threads to another spot, release the old filaments, and pull itself to the new site—it is a creature quite unlike the sedentary oyster. Also, unlike the oyster, the mussel uses two muscles to close its shell, one at each end of its body. It also has a single short siphon that is used only for exhaling waste; the mussel draws nutrient-laden water in through the slit between its valves. The shell itself is neither rough nor irregular but somewhat smooth and patterned.

Some marine varieties of mussel are quite edible; in fact, thousands of tons are eaten yearly in Europe. Other mussels, principally those in Europe and North America, produce pearls. At one time the freshwater pearl mussel, an excellent pearl producer, was carefully managed in both locations. Now, however, water pollution and the cultured pearl market have pushed the industry off the market.

As with the oyster, the scallop (family Pectinidae), with its symmetrical shell (famous as the Shell Oil Corporation logo), has only one muscle with which to operate its valves. But like the mussel, it uses its "foot" to crawl along rocks. Unique to the scallop and to its brother the file-shell is its ability to propel itself forward or backward through the water by quickly squeezing water through the openings near either its hinge side or its front—it has no siphons. This squirt is so powerful that scallops have been seen to "jump" out of the water onto land. Although it can "swim" around underwater, some, such as the southern scallop that lives on North America's Pacific coast, prefer a different mode of travel. It walks along the sea bottom on its foot. Like the mussel, when some scallop species find a suitable place to lodge, they will tie themselves down with excreted threads. Others, just like the oyster, permanently cement one of their valves to a rock or another solid foundation.

What really sets the scallop apart from other shellfish is its hundreds of eyes. These organs, located around the edge of both shells, detect the motion of predators such as starfish and octopi. If by accident any of the eyes are lost or damaged, new ones grow in as replacements.

Flounder ● Halibut ● Turbot ● Sole

By definition, a taxonomic group of animals is united by its shared features. Yet it is not uncommon that such a group includes members with quite deviant or exotic characteristics. Within the vast world of fish, there is no shortage of such aberrant members. And such is the case with the quartet of fish in question, commonly called flatfish. Although their odd body form borders on the bizarre, it is clearly geared toward survival, allowing them to rest on the sea bottom nearly undetected.

Flatfish start out in life appearing quite normal, yet soon after birth a strange phenomenon occurs: their heads and their bodies inexplicably start to rearrange themselves. The skull bones, nerves, muscles, and skin all begin undergoing complex alterations, resulting in a very curious transformation. The flatfish's body flattens from side to side like a pancake—one side slightly arched—and tilts over until the flatter side is facing down. This side is seldom well pigmented and in some species bears smooth scales, whereas the scales on the upper side become toothed. In addition, the pectoral fin on the underside seldom fully develops, yet the dorsal fin keeps growing until its base extends over the head. Also, the dentition of some may alter, the mouths becoming distorted in an attempt to remain horizontal. But the most remarkable change in this two- to four-month-long transformation is the loss of the underside eye. The eye does not disappear but moves across the face to join the upward-facing eye. In most species the migrating eye is always the same. For others, either the left or the right eye may stray, depending on which side is uppermost. The species are designated either righteye or lefteye flatfish, depending on which side both eyes occur.

During this transitional period, other changes are also taking place.

As with the eyes, the underside nostril shifts to the upper side. If any are born with swim bladders, these are soon lost, compelling them, like the majority of flatfish, to spend most of their time on or near the sea bottom. Also, color-adaptive skin cells develop that can alter the pattern of the darker upper side to match the background of the fish's habitat. This ability is so well developed that in laboratory tests, some species were able to reasonably mimic the checkered patterns on which they were placed in 2 to 8 seconds.

The flatfish order, Pleuronectiformes (meaning "side swimmer"), consists of 6 or 7 families (authorities differ on the number), including 2 of flounder and 1 of sole. Those flounder whose eggs contain a single globule of oil in their yolk usually bear lefteye species and form the Bothidae family. Flounder without such an oil globule most often produce righteye species and make up the Pleuronectidae family. Sole, forming the family Soleidae, are always right eyed, their yolks containing numerous oil droplets.

The term "flounder" comes from the Norwegian *flundra* and in Great Britain is often reserved for one particular species from Pleuronectidae: *Platichthys flesus*. The North American and Continental European use of "flounder" is more encompassing and covers several species of Bothidae and Pleuronectidae, some of which bear more specific names. Of these, dab and plaice species are very popular in Europe, as are halibut and turbot, fish also familiar to North Americans.

Those flatfish North Americans call "halibut" can be found in both flounder families; the California halibut is a predominantly lefteye species, and the Pacific and Atlantic halibut, a righteye species. The name "halibut," meaning "holy turbot," was originally applied to those righteye flounder eaten on holy days in Scandinavia. Today, commercial fishermen still favor the righteye halibut. These plump, dark-colored flatfish often reach 500 pounds or more, with the record holder having attained 700 pounds and 15 feet in length. Most flatfish are far smaller, many weighing less than 10 pounds. Of all the flatfish, the halibut changes the least during their transformation. Its large mouth remains uncontorted, and in some species the migrating eye makes it only to the center of the head. Unlike other flatfish that find much of their food on or near the bottom, the voracious halibut is a powerful swimmer that does not hesitate to pursue prey into open waters.

The turbot, a member of the Bothidae family, is reportedly the best tasting of all the flatfish—some say of any fish in the world. These large-mouthed, diamond-shaped fish average thirty pounds and reach forty inches in length. Their scaleless skin is generally light colored on both sides, with a scattering of tubercles on the upper surface. Turbot, whose territory ranges from the Mediterranean to the North Sea, are relatively shallow sea feeders, seldom swimming below 250 feet.

Challenging the turbot in taste is the sole, although to attain optimum flavor, this smaller flatfish must lie dead for two to three days before cooking preparations are begun. The European sole, *Solea solea,* won such widespread distinction as "fillet of sole" that this culinary term is now casually attached to many other less tasty flatfish. Such exploitation has even led to the misuse of "sole" for species from both families of flounder, such as the lemon sole and the American Pacific fantail sole from Bothidae. If one intends on having true sole for dinner, the caveat "buyer beware" would seem prudent.

Sole, reportedly so named because of their resemblance to the sole of the human foot, inhabit marine and fresh waters. Some favor the cooler waters of Europe and the Americas, but most prefer warmer seas such as those between Australia and Japan. Like many flatfish, most sole live in shallow waters, yet a few can be found at depths of up to 3,800 feet.

Unlike the other flatfish, the sole has no ribs, nor does it have more than ten abdominal vertebrae—the flounder always has ten or more. More apparent are the many external differences that set the sole apart from the other flatfish, including its size, which is usually less than a foot in length and only rarely exceeds two feet. Also of note is how its dorsal, pelvic, anal, and sometimes tail fins run together, forming an almost continuous ribbon of fin around its teardrop-shaped body. Sole eyes are quite small, close together, and, being somewhat underdeveloped, less movable than those of other flatfish. On most fish, including the flounder, a discernible fold or a delineated ridge called the preopercle sits in front of the gill cover, indicating the rim of the cheekbone. On the sole, this is well hidden by the skin. To the rear of the sole's gill cover and extending to the tail runs a straight row of tiny pore-like openings called the lateral line. On the flounder, this line is not entirely straight but begins with a hump or a short diagonal segment from above.

Herring ● Sardine ● Pilchard ● Anchovy

For hundreds of years, the herring and sardine fishing industry supported numerous villages along the coasts of northern Europe, Alaska, Siberia, and Japan. Up until the 1930s, towns such as Monterey, California, thrived on the commerce these vast shoals of fish provided. But, eventually, overfishing and wastefulness took their toll, decimating the world's supply of herring and sardines and leaving behind a collapsed fishing industry. To meet the demand for the small silvery fish, anchovy fishing off Peru quickly grew. But by the 1980s, it, too, failed for the same reasons: imprudent policies and unenforced regulations. The worldwide appetite for these fish has left us with a multitude of other names, in addition to the terms "herring" and "sardine"; these include pilchard, sprat, cailleu, bang, lapad, maray, tamban, blueback, black belly, digby chick, and smid.

The word "herring," emerging centuries ago from West German languages, has gone through various spellings but is traceable to the Anglican *hering* of A.D. 700. Thought to be related to the Old English *har,* for the color gray, it was applied to the most plentiful fish species of its kind in the area, the silver-gray *Clupea harengus,* which is still found in limited quantities along the European coasts of the North Atlantic and as far south as Cape Hatteras. "Sardine," a more recent term, dates from about 1400, when it was spelled *sardyn,* and reflects the source from which it is derived: the waters surrounding the Mediterranean island of Sardinia. Principally a Mediterranean fish, the specific species once called the sardyn, *Sardina pilchardus,* is somewhat similar to the herring but is noticeably smaller.

"Pilchard," a term of obscure origin and found in the species name of the "true" sardine, is frequently used in place of "sardine." Europeans commonly apply "pilchard" to the mature *S. pilchardus* of 9 to 10 inches and reserve "sardine" for the juvenile of 5 to 6½ inches. The British even consider "pilchard" to be the fish's proper name and often use "sardine" as a commercial designation when the fish is dried, salted, smoked, canned, used for its oil, or used as meal. Yet they also use "sardine" for several species of other genera, regardless of their commercial status. The French use "pilchard" to denote sardines or herring that are canned in tomato sauce or oil.

Sometimes such an extensive focus on a specific commodity of nature—in this case, the demand for a particular food fish—leads to the abuse of a term until its definition is so enlarged and often corrupted that its original meaning, if not rendered ambiguous, is all but forgotten. This is the dilemma that has befallen "sardine"—confusion brought on in part by the mixed British use of the word. When "sardine" was adopted for the smaller, more cannable pilchard, its use eventually grew to encompass similar small fish. So, not only were the sardyns (the young pilchards of the British) "sardines," but so, too, were any other small cannable herring, and a number of other herringlike fish of the same order. This questionable use eventually led to such widespread misunderstanding that as a commercial term, its definition has now been left to the discretion of individual countries.

In the United States the Food and Drug Administration publishes an Internet-accessible guide entitled "The Seafood List" (previously called "The Fish List, a Guide to Acceptable Market Names for Food Fish Sold in Interstate Commerce" in its print form). The guide lists four types of names for each species of food fish: (1) its *scientific* name; (2) the preferable *market* name; (3) one acceptable *common* name, descriptive of the species; and (4) various *regional* names whose uses are discouraged. Sixteen species within four genera go by the market name "sardine." Of these, "pilchard" appears as an alternative to seven of them. The common name of each sardine species includes the term "sardine" in all but seven cases, where "pilchard" is substituted. "Herring" appears as a market name for 15 species contained within 8 genera. As a common name, it shows up in 8 species. Regional names seem to have been bestowed quite arbitrarily for both herring and sardines. "Sardine," "pilchard," "sprat," and many other names all appear in various combinations for different specific species. The red-ear sardine, for instance, goes by the regional names of sardine, sardina, sprat, pilchard, and redear pilchard. "Sardine" is also used as a regional name for a species of anchovy. The 21 species of anchovies sold in the United States all go by the market and the common name of "anchovy." Only the Japanese anchovy has an ill-conceived regional name: halfmouthed sardine.

Although regional and vernacular names are discouraged, fish distributors, among others, routinely ignore this admonition; and other countries may have very different ideas about these fish and what they should be called and have exported them to United States markets under nonstandard names. Unfortunately, this varied use of the terms "herring" and "sardine" makes it difficult to construct a useful definition of either or to make much of a distinction between them, although it has prompted some ill-conceived attempts. At the extreme are the remarks by a few science popularizers, denying the existence of any such fish as a sardine, explaining that fish (primarily, herring) become sardines when they are packed in a sardine can!

So, inasmuch as "sardine" seems to have been purged of any universal or practical meaning, and with "herring" faring only somewhat better, can nothing be said to distinguish the two? Well, it can be helpful to note the characteristics of the families involved and to examine the most prominent or archetypical species denoted by each term. For the herring, this would be the Atlantic herring, *Clupea harengus,* and for the sardine, the "true" sardine, *Sardina pilchardus.*

Taxonomically, herring, sardines, and anchovies all belong to a large order of fish (Clupeiformes), which also includes the tarpon, the trout, and the salmon. The three families containing herring are popularly labeled herring (Clupeidae), wolf herring (Chirocentridae), and denticle herring (Denticipitidae). Although both the herring and the two wolf herring species are commercially important, only the former is listed in the FDA guide. Wolf herring average close to 3 feet in length and have large fang-like teeth. The lone denticle herring is a 3-inch long African freshwater species, with many small toothlike growths on its head and the fore part of its body. Clupeidae, besides harboring the herring, also includes sardines, sprats, shad, and alewifes, so its popular designation as "the herring family" is one of convenience, and no doubt one reason sardines are often thought of as a type of herring. The 21 species of anchovies, which occupy 5 genera and are closely related to the herring and the sardines, belong to the family Engraulididae.

The Clupeidaes (216 species in 66 genera) are typically a small-mouthed, silvery-sided fish, generally less than 18 inches long, which has a row of sawtoothed scutes (bony projections) on its belly. Its tail is deeply forked, and its lone dorsal fin sits in the middle of the back. All have very oily flesh, usually covered by large, thin scales. Whereas a few live in brackish or fresh water, most inhabit a wide range of marine waters, usually swimming in large schools. Unlike the few deep-swimming species of the icy northern waters, most Clupeidaes are found in shallow waters, including the warm seas of the tropics.

Herring, although comprising only a fraction of their family's many genera, have the distinction of having their name selected as the informal designation for the entire family, Clupeidae. This honor, no doubt accorded because of the herring's early commercial value, rests primarily upon the species *Clupea harengus,* the Atlantic herring. The only other species in the genus is *C. pallasi,* a slightly smaller fish commonly called the Pacific herring (some authorities regard it as a subspecies of the Atlantic herring). The somewhat deep-bodied profiles of these herring and their length (15 inches) are the most readily noticed features distinguishing them from the true sardines (9 inches). Also, in contrast to the sardine, the herring's belly scuts lack a sharp keel, and it has no spots on its body. More important, however, the Atlantic herring is taxonomically set apart from the "true" sardines by its anal fin and smooth gill cover.

The "true" sardine has gill covers with three to five striations or ridges radiating toward the rear. Also, the last two rays of its anal fin are elongated, a characteristic not found on herring. In general, this small fish has a round, spindle-shaped body, covered by large loosely attached scales, and usually one or more spots to the rear of its gill cover. Such features identify not only *Sardina pilchardus,* the first recognized so-called true sardine, but also most sardines in the other four, fairly well-settled, genera, many of which were previously classified in *Sardina.*

The anchovy is thinner, rounder, and smaller (seldom over 8 inches long) than either herring or sardines. Most noticeable in this silvery fish is its prominent snout, which hangs over a rearward-set lower jaw, imparting a somewhat chinless appearance. The mouth itself is very large, extending in some species back beyond the gill cover. Of the many anchovy species, only those of the North Atlantic carry belly scuts. All swim in schools, most sticking to coastal waters, but a few can be found in the fresh river waters of the Amazon and Southeast Asia. Many of those caught commercially in the Pacific are used as tuna bait, whereas the European species often end up in sauces or pastes.

Lobster • Crayfish • Crawfish • Scampi

These fresh- and saltwater creatures, highly valued as dinner fare, are close relatives of commercial shrimps and prawns: all belong to the order Decopoda (meaning "10 legs"). Where they differ most noticeably from shrimps, other than their size, is in their mode of travel. Most shrimps get around by swimming, whereas lobsters and their kin usually prefer to crawl. Focusing on this characteristic, zoologists divide the order in two, placing shrimps in the suborder Natantia and lobsters in the suborder Reptantia, along with the crabs.

Lobsters are carnivorous marine animals that come in two varieties: those with claws, often called true lobsters (infraorder Astacidea), and those without claws, the spiny lobsters (infraorder Palinura). There is a third variety of reptant with claws, sometimes called a mud lobster; however, because of certain unique features (for one thing, it has a soft abdomen), it does not fit in with the true lobsters and makes up its own infraorder, Thalassinidea.

The true lobster (the term "true" here has no scientific significance), sporting a pair of formidable claws, one designed for crushing and the other for grasping and tearing, represents the epitome of defensive and offensive development in crustaceans. Because the Norwegian and South African Cape true lobsters are somewhat small creatures, they are not particularly formidable, but the European and American species, near-Goliaths of the crustacean world, can be a true challenge to any predator. The largest lobster ever captured was an American *(Homarus ameri-*

canus) that in 1977 weighed 41 pounds and reached 3½ feet in length (the giant spider crab with outstretched legs spanning up to 12½ feet is essentially bigger, but with a body only 15 inches wide, it weighs no more).

The spiny lobsters (some of which are called langouste, and rock, slipper, and Spanish lobsters) are a bit smaller—the largest ever brought in weighed just over 31 pounds. Because they lack claws for defense, they have developed unique spine-encrusted antennae, which are much stronger than those of the true lobsters and serve as effective whips to ward off adversaries. As further protection, they also bear numerous spines on their carapaces (enveloping shells). The spiny lobster differs from the true lobster in the construction of its tail fan. The uropods, which are broad segments that fan out to the sides of the tail's rear, are divided into two parts on the true lobster but remain undivided on the spiny lobster. As a seafood, the spiny lobster lacks the meaty claws and legs of true lobsters, yet because of its particularly tasty tail, it is still highly valued.

Three species of true lobsters account for almost all commercially harvested varieties: the American and the European, which inhabit rocky bottoms and crevices, and the Norwegian, which favors deeper waters and muddy bottoms. The Norwegian lobster, *Nephrops norvegicus,* is a pale species about 8 inches in length and armed with long, slender pincers. It is found in the Mediterranean Sea and northward into the waters surrounding the British Isles, where English fish merchants market it under the erroneous name of Dublin Bay prawn. Italians who harvest it from the Adriatic Sea also call it scampi. "Scampi," actually the plural of "scampo," is often misused for shrimps broiled in butter and garlic.

Crayfish, unlike lobsters, are omnivores. Strictly freshwater crustaceans, they feed not only on mollusks, worms, and fish but also on plant material. Comprising more than 500 species, they can be found throughout the world, except for the African continent and most of the Orient—they do occur in Korea and northern Japan. The majority lives in the shallow waters of ponds, lakes, streams, and, oddly enough, caves. Living in caves where light is absent, these crustaceans, like other deep-cave dwellers, have lost all pigment and eyesight; their eyes have completely disappeared.

The largest crayfish in the world inhabit the small streams of Tasmania. These reach 2 feet in length and weigh 8 to 9 pounds, quite a bit larger than the average crayfish of 4 inches. Except for their size, crayfish closely resemble true lobsters and, accordingly, belong to the same infraorder, Astacidea. Most are found in temperate zones, those of the southern hemisphere differing from their northern relatives by the lack of appendages on their first abdominal segments. Although generally regarded as simply small freshwater versions of the clawed lobster, crayfish, in fact, can be distinguished from lobsters by two important characteristics: their form when hatched and the maintenance of salt equilibrium in body fluids.

After a young lobster hatches, it goes through a series of larval stages until it eventually resembles an adult. This transitional period is called the zoea stage of development. In contrast, the crayfish undergoes this stage while in the egg; subsequently, when it hatches, it already resembles its adult form.

As marine invertebrates, lobsters easily maintain a proper concentration of salt in their body fluids. Water taken in through their gills contains just enough salt to balance the essential body concentration and prevent dilution. Crayfish, freshwater inhabitants, do not have the luxury of a salt-balanced environment. For the crayfish, fresh water is continually being taken in and diluting the body's salt equilibrium. To counteract this, it excretes excess water not only from its excretory pore, as do lobsters, but also through gill surfaces. In addition, it reabsorbs any salt destined for the excretory pore through a convoluted tubule located just before the bladder.

"Crawfish," like "crawdad," is a corruption of "crayfish." The two terms are particularly popular in the southeastern United States, where crayfish dishes are favored. Oddly enough, "crawfish" has also been used to identify the spiny lobster. London fishmongers and residents along the coast of Great Britain commonly use the term in this fashion. Moreover, some British authorities condone this use as a valid label for the spiny form. Why a term that is a variation of a name for a particular form and classification of animal (clawed freshwater Astacideans) is used for a quite different one (unclawed marine Palinuridae) is not clear. No less perplexing is the use of "marine crayfish" for the spiny lobster, so as to differentiate it from the crayfish. This commonly occurs in places such as Australia, where true lobsters are absent or not marketed. Seemingly, "lobster" would be a better choice than "marine crayfish."

Marlin ● Swordfish ● Sailfish ● Spearfish

Of all the blue-water game fish in our seas, these four long-billed creatures are considered the elite, their sleek beauty and exceptional fighting ability making them the premier quarry of the deep-sea angler. From the smallest, the spearfish (53–154 pounds), to the largest, the marlin (181–1,997 pounds), all put up a challenging fight. Yet the ultimate contest between man and fish comes not from the largest of these but from the swordfish (up to 1,182 pounds), particularly any over their average weight of 250 pounds. With tender, easily torn mouths, the swordfish's tactic of powerful runs and repeated leaps into the air, plus its penchant for taking short rests to revitalize itself, continually puts an angler's skill to the test—a confrontation that can last 3 to 4 hours. The remaining member of this quartet, the sailfish, weighs in at between 130 and 223 pounds.

To the sport fisherman, all four are thought of as billfish, a reference to a billfish's most prominent feature: an elongated spear- or swordlike projection on the front of its head (an extension of its premaxillae and nasal bones). As fitting as the term "billfish" may be, scientists are a bit more restrictive in their use of the word. Although the term is a popular, rather than a scientific one, scientists often limit its use to those fish of the Isiophoridae family: marlin, sailfish, and spearfish. This usage is also supported by the Fisheries Management Councils of the United States. The swordfish, with its uniquely shaped bill, is the lone species in the family Xiphiidae; and whereas its bill does qualify the swordfish as a billfish, ichthyologists (fish scientists) prefer to call it a "broadbill."

Of the four, the swordfish sports the longest bill, which ranges from 20 to 33 percent of its total body length. Unlike the rounded bills of the others, that of the swordfish is flattened horizontally into an oval cross-section (hence its nickname, "broadbill"), with a black-tinted top and a pale underside. One of its functions is to cut through the water while swimming, enabling the swordfish to reach speeds up to 60 mph. As for impaling prey, this seems to happen by chance, rather than through design. The swordfish catches most of its food by snapping it up with its mouth. Only when it encounters a large school of fish does it flail its sword around, slashing from side to side, stunning and occasionally spearing fish. The other billfish use this tactic as well, some even pursuing shoals of fish for days in a recurring cycle of chase-eat-rest.

In contrast to its oval bill, the body of the swordfish is cylindrical, whereas the bodies of the other billfish, all of which have round bills, are oval. This difference in body shape allows a swordfish to carry up to one-third more body mass than, say, a marlin of equal length. Fittingly, some of this extra bulk is muscle, producing a very powerful body with exceptional endurance. But these are not the only features that distinguish the swordfish. Although born with strange, spiny scales, these are soon lost, leaving the adult broadbill scaleless. The swordfish also lacks the pelvic fins (those long, thin appendages that trail loosely from beneath the breast) that are borne by the other billfish. Its first dorsal (top) fin closely resembles that of the shark, its base generally extending back no farther than does its top. Marlin, sailfish, and spearfish all carry their dorsal fins down their backs to at least the point where their anal fins begin. From the anal fin to the tail fin is the "neck" of the tail, or the caudal peduncle. On each side of this "neck," the swordfish carries a single keel-like structure, whereas the other billfish carry a pair of such keels. And unlike the distinctly rounded foreheads of the other three, the forehead of the swordfish remains fairly flat all the way back to the dorsal fin. The last telling feature of the swordfish is its eyes, which are proportionally much larger than the eyes of the other three.

Internally, the swordfish possesses three notable features. One, which is extremely unusual, is a set of muscles situated beneath the brain that contains "heating cells." These provide warmth to the brain and the eyes, permitting the swordfish to pursue prey into near-freezing waters. Second, the swordfish has no teeth in its jaws; the other billfish, often reported as toothless as well, do, in fact, develop extremely small teeth no larger than sand grains. Finally, and taxonomically very important, the swordfish has 26 vertebrae, whereas the other three have only 24.

For the sailfish, the marlin, and the spearfish, the most distinguishing feature is their first dorsal fin. The sailfish, aptly named for this feature, sports an enormous spotted blue fin on its back. It tapers very little as it extends backward and is much higher than the body of the fish is deep. A potential hindrance to high-speed swimming, the fin folds neatly into a groove along its back, enabling the sailfish to match the swordfish in speed. Although the sailfish is not as indefatigable a fighter as the swordfish is, it appeals to the sport fisherman because of its fighting spirit and dazzling aerial leaps. The sailfish is not a particularly varied fish; its genus contains only one species or two, if one separates those of the Atlantic from the Pacific form (authorities disagree on this point).

Marlin classification is also unsettled with species assigned to two genera, *Makaira* and *Tetrapturus*. Because *Tetrapturus* is the only genus of spearfish and therefore considered the "spearfish genus," *Makaira* is usually considered to be the "true" marlin genus. In the past the marlin of *Tetrapturus,* the Pacific striped and the white, were often listed in *Makaira,* but, based on vertebrae position and preadult bill size, they were transferred to the spearfish genus. Only three species are commonly recognized within *Makaira:* one black and two blue marlin. However, some authorities separate out the cape marlin and, in some taxonomies, include the short-nosed spearfish. Unfortunately, this unsettled and double-genera classification of marlin raises questions as to the validity of their defining features.

As with sailfish, the dorsal fin on the marlin and on the spearfish also folds into a groove on the back. In the marlin, this fin reaches only the height of its body depth in the white species, but in both species it quickly tapers in size before trailing down the back in a low profile and ending just short of the second dorsal fin. For sport-fishing records, marlin are separated into five categories: black, striped, white, and two blue. The blue is a huge billfish, weighing close to a ton in record catches; the biggest white marlin caught weighed only 182 pounds. So, unlike the others, the range in size of the various marlin species is extremely wide.

As for the spearfish itself, the trailing portion of its first dorsal fin is much taller than that of the marlin and in some species remains higher than its body depth for most of its length. Its slender, lightweight body and a bill that is the shortest of all the billfish can also distinguish the spearfish. In the long-billed species, the spear may reach twice the length of the lower bill, whereas in the short-billed species the spear is scarcely longer than its lower jaw. With the exception of spearfish, the anal vents of billfish normally sit just in front of the anal fins. On spearfish, this vent is located well ahead of the anal fin. Unlike the other billfish, these small, open-water fish are not great fighters. Consequently, they are not a favored quarry. Yet when taken with light tackle, they can be fine opponents.

Minnows ● Suckers ● Other Small Fish

Among avid fishermen, it is not uncommon to hear "minnow" used for a wide variety of small freshwater fish, a practice that appears to be grounded in the notion that the term refers to either the young of most any fish or simply small bait fish in general. However, properly used, "minnow" is better limited to specific fish of the Cyprinidae family. This varied family of over 2,100 species contains not only the small bait-size minnow species but also goldfish and carp.

Because minnows are the only members of Cyprinidae native to North America, the family is often referred to as the minnow family. Unfortunately, this has helped foster the idea that goldfish and carp are minnow species, despite the distinguishing features of the two. And when one considers that "minnow" comes from the Anglo-Saxon *myne*, meaning "small," it is truly an inappropriate label for carp, almost all of which are significantly larger species, some reaching 9 feet in length.

Although Cyprinidae is the largest family of fish in the world, its members are completely absent in South America and Australia. In the United States and Canada, 231 native minnow species, plus various introduced carp and goldfish, represent the family. But it is China, from which our goldfish and carp originated, that is home to the most Cyprinidaes—nearly 600 species.

Many of the small fish that merit the name "minnow" are also known as chub, dace, shiner, splittail, stoneroller, and squawfish. Some terms, such as stoneroller and squawfish, are limited to a single genus, whereas others, such as chub, dace, and shiner, cover several genera. Each of these colorfully named groups is generally organized around some particular feature, features also found in fish of other families. Unfortunately, several of these lookalikes from other families are sometimes erroneously called minnows—specifically, the pin minnow (a sucker), the Lake Eustis minnow, the sheepshead minnow, and the topminnow (all killifish), plus one entire family, the mudminnows.

The characteristics that distinguish minnows and their kin are not unique, but, taken together, they describe only Cyprinidae. Most notable are the rays in their fins and the lack of teeth in their jaws. Minnow rays are bony, segmented, and often branched projections—the dorsal fins usually bearing eight or nine. These are normally somewhat flexible, but on goldfish and a few others one or two rays may harden like spines and have been erroneously called such. (Spines are unsegmented and unpaired.)

A minnow carries all its teeth in one to three rows on the last gill arch in its throat. Variations in this feature are important for species identification, although to examine these pharyngeal teeth, one must first dissect the fish. Other significant minnow features include a bare head (except for those mature males that develop tubercles during the breeding season); a single dorsal fin; no adipose fin (a rayless back fin located between the dorsal fin and tail); usually smooth, thin lips; and cycloid scales (those with a smooth surface).

Suckers, particularly the smaller species, are frequently mistaken for minnows. As fairly close relatives of the minnow (they belong to the Catostomidae family), they possess several minnow characteristics, including pharyngeal teeth, soft-rayed fins, naked heads, and cycloid scales. Where they differ most noticeably is in the construction of their mouths. The sucker mouth is usually small and ringed by large and heavily creased or folded lips. This enables the fish to extend its mouth and draw in food from the lake or stream bottom. Another telling feature is the position of the anal fin. Suckers can be easily distinguished from native North American minnows by applying a simple test called the anal fin rule, which states: If the distance from the nose to the beginning of the anal fin is more than two and one-half times the distance from that anal fin point to the tale fin, the fish is a sucker—if shorter, it is a minnow. Not quite as reliable is the observation that a sucker usually has more than nine dorsal rays, the maximum minnow number. And no sucker has hardened spine-like rays. Finally, a sucker carries only one row of teeth on its last gill arch, as do some minnows, but unlike such minnows, which bear only five teeth, the sucker possesses many more.

Those minnows named killifish, a few of which do somewhat resemble the Cyprinidaes, can be dismissed as minnows (and suckers) if one looks at their teeth. Either cusped or pointed, they are easily found lining their jaws. Also, killifish have scales on their heads, and their tails are not forked like those of minnows but rather are squared or rounded. These same observations can be made of the mudminnows, a small family of four American species that makes its home in muddied waters. Among other features, the mudminnow can be distinguished from the killifish by the teeth it carries on the forward portion of its mouth roof, teeth that killifish lack.

Sharks ● Other Fish

Few animals are as well known by deprecating clichés as the shark. The creature is universally regarded as a fearless predator, a creature of savage grace, a voracious man killer, a superbly designed killing machine, and a surviving relic of the distant past, driven by the basest instincts—impressions reflecting the macabre fascination these animals invoke in us. And those species exhibiting a docile and even friendly disposition do not seem to temper these perceptions—of the 368 species studied, only 21 have been known to attack humans. The shark shares its infamous status with no other. It is a unique animal, not only in behavior but in anatomy as well.

Most people readily perceive the shark to be anatomically quite different from other fish: its underslung jaw with large serrated teeth and a streamlined body, betrayed by a surface-slicing dorsal fin, all spell "shark." For many years science had recognized the shark's distinctive features, but it was not until around 1910 that sharks were separated from the other fish to form, in part, a new class. Today, they share this class, Chondrichthyes (meaning "cartilaginous fishes"), with the skates, the rays, and the chimaeras. Except for the eel-like lampreys and hagfish (excluded because they lack jaws) and also the lungfish (in some classification schemes), all other fish belong to the class Osteichthyes: the bony fish. These class names reflect the principal difference between sharks and the other fish: the skeleton of the shark is composed of cartilage, rather than of true bone. Although almost all Osteichthyes, such as marlin, cod, trout, perch, herring, and so on, have skeletons made entirely of bone, a few barely merit inclusion in the class. Sturgeon, for example, which have been called "living fossils," have skeletons consisting almost exclusively of cartilage.

Cartilage is the flexible, translucent gristle that supports the end of our nose and shapes our ears. It is composed of connective tissue components in a matrix of protein and, as with the Chondrichthyes, is often strengthened with calcium. Although a bony skeleton may seem evolutionarily more advanced and therefore superior, fossil evidence suggests that sharks and bony fish evolved simultaneously from a common ancestor. A bony skeleton does have certain advantages, such as providing stiff structural support for the body and the fins, but for the shark, a skeleton of cartilage provides two important benefits. Its elasticity gives the shark greater flexibility than does bone and therefore more maneuverability in pursuing prey, and because the shark does not have a gas-filled swim bladder to keep it buoyant, as do most bony fish, the lower density of cartilage helps to prevent it from sinking.

The skin of sharks is also very different from that of bony fish. Except for those few fish having no scales, such as catfish, bony fish carry thin, overlapping scales of varying sizes. Distinguished by their shape, these plates consist of bony material that forms out of the dermal layer of skin and develops annular rings as the plates grow. Unlike shark scales, which "cut the gum," those of bony fish retain a thin covering of skin tissue. Shark scales, or dermal denticles, are quite different and come in many forms. Most are streamlined, with grooves, flutes, or ridges running parallel to the body length—an aid in reducing swimming drag. These scales do not grow as the shark ages but are continually replaced by larger denticles. And unlike the simply constructed scales of bony fish, those of sharks, oddly enough, closely resemble the structure of teeth. Developing from dermal tissue, denticles are composed of an inner pulp cavity that contains blood vessels and nerves, which is surrounded by a dentine layer with an outer covering of enamel derived from the epidermis (top layer of skin).

Unlike the shark's skeleton of cartilage, shark teeth, with their tough enamel coating, are usually the only parts that survive fossilization. Not set directly into the bone of the shark's jaws, as with many other fish, shark teeth develop directly out of the gums. They are, in fact, modified forms of the shark's denticles, which may have evolved ages ago from scales that migrated to the mouth. In contrast to shark teeth, which line only the rim of the jaws, the teeth of a bony fish, depending on the species, appear in a number of places: in the jaw bone, in the sides and the roof of the mouth, on the tongue, or in the throat. Most shark teeth, in their varied shapes, are typically pointed; however, a few sharks possess flat, pavement-like teeth, used for crushing. And unlike the irreplaceable teeth of bony fish, old or lost shark teeth are soon replaced by new ones that migrate up the jaw and into position. These budding reserve teeth are so numerous that in some species, as many as twenty rows of them may be slowly inching forward.

Surrounding the shark's dental work is a skull composed of three principal parts: a seamless cranium (bony fish have sutures in theirs); usually, a shortened lower jaw, with tremendous crushing power; and an upper jaw suspended from the cranium by ligaments (not fused, as in bony fish), which in feeding is thrust forward to enlarge the mouth.

Along the side of the head and connected to the mouth of fish are the gills, which take oxygen out of the water and release carbon dioxide waste. The shark has five to seven on a side (those with more than five gills are considered to be more primitive species), each separated from the other by a partition that extends to the outside of the body, forming a skin flap. As the shark moves through water, these flaps easily flex closed, giving the gills time to absorb the oxygen. Although it may not appear so, bony fish have four pairs of gills, none of which carry partitions. Instead of flaps to regulate the flow of water, a bony fish has a rigid cover, called the operculum, that opens and closes and shields the entire opening. The single opening on the side of a bony fish versus the five to seven on a shark is a sure way to tell the two forms apart.

Sharks bear several other distinguishing features, some minor, and some typical but not exclusive. The following are seven of these more notable features:

- Lacking swim bladders to help keep them from sinking, most sharks have enormous livers filled with low-density oil, giving them much of their buoyancy. In some species, these organs account for up to one-third of their body weight. Bottom-dwelling species have proportionally smaller livers than do their more ocean-going relatives. Bony fish have normal-size livers—even those without swim bladders, such as the billfish and the tuna.

- The pelvic fins on male sharks bear clasping appendages that assist in transferring sperm during mating, effecting internal fertilization. Only a few groups of bony fish practice internal fertilization, and this is accomplished by using a modified anal fin, not pelvic claspers. Bony fish can also rotate their pectoral fins, an ability never developed in sharks.

- Most sharks bear live young, whereas almost all other fish lay eggs. Guppies, mollies, and swordtails (small fish often kept in aquariums) are among the live-bearing exceptions.

- The eggs from those sharks that do not deliver live young are usually encased in horny capsules, some spiral in shape and others more purselike. Bony fish do not lay encased eggs.

- The number of young or eggs produced by sharks is small, ranging from only a couple in some species to about 165 in the blue shark. Bony fish are much more prolific. The ocean sunfish lays around 30 million eggs a year, and one louvar (a deep-sea fish) had an estimated 47.5 million eggs inside her when caught.

- The vertebral column in a shark ends in the upper lobe of the tail, which is often much longer than the lower lobe. In most other fish the vertebral column effectively ends at the base of the tail, and both lobes are typically symmetrical.

- The fairly short ribs of a shark do not protectively enclose its internal organs, as do those of most other fish.

If, in considering these differences, you are left with the impression that sharks seem somewhat less than qualified to bear the label "fish," you are not alone. Richard Ellis, in his fascinating *The Book of Sharks,* addresses this thought and remarks, "In fact, the differences are so numerous that some would argue sharks are more closely related to mammals than they are to bony fish."[1] This may very well be true, but if so, would it still be proper to call them fish? Probably not, but as Ellis observes, "calling something a fish does not necessarily make it a fish."[2]

Shrimp ● Prawn ● Krill

Unfortunately, over the years the original meanings of "shrimp," "prawn," and "krill" have become somewhat corrupted, if not altogether lost; the three terms, lacking any scientific standing, are now often used interchangeably, particularly "shrimp" and "prawn."

"Shrimp" (originally meaning "to contract or shrink") is in fact a label bestowed on an assortment of crustaceans, many of which look nothing at all like those we serve as seafood. The mantis shrimp (order Stomatopoda), for instance, can grow to a foot in length; it has a narrow body and a broad tail and, instead of pincers, uses claws armed with sharp spines. Another truly odd-looking crustacean is the skeleton shrimp (order Amphipoda), which somewhat resembles a slender scorpion tail with short legs. But the strangest of all is the seed shrimp (subclass Ostracoda). This tiny ($^1/16$–$^1/8$ inch long) creature lives within a bivalved carapace, or "shell," giving the appearance of a tiny fringed bean. Almost as strange are the brine shrimp, the clam shrimp, the fairy shrimp, and the tadpole shrimp: all "shrimps" of other orders.

"True" shrimps, or what we customarily think of when hearing the word, belong to the Decapoda order of the subclass Malacostra, the order that also contains crayfish, lobsters, and crabs. Although the designation Decapoda implies ten legs, in many species one or more pairs are modified into pincers and may no longer function as walking legs. The common pink shrimp is typical, bearing pincers on its first three pairs of "legs." Even within Decapoda itself, shrimps vary greatly in appearance. The delicate one-inch-long transparent Pederson's cleaning shrimp is quite unlike the edible ten-inch pink shrimp or the snapping shrimp, which bears a huge claw that makes a loud snapping sound.

Each of the shrimps of Decapoda has a somewhat long cylindrical body, with longitudinal abdominal segments. It also has a fused thorax and a head that is protected by a carapace. The carapace is a dorsal shell that extends down the sides and encloses the gills and, in many species, also extends forward above the eyes to form a structure called the rostrum. The eyes, like those of lobsters and crabs, are borne on stalks. Every decapod shrimp has a tail fan composed of appendages called the telson and uropods. The limbs often grow quite long, with the first three pairs modified to serve as mouthparts. The antennae of many shrimps are also elongated, which in some species reach more than twice their body length. Of all the various species, the larger and commercially important species such as the pink shrimp are of concern here.

Hundreds of years ago, the term "shrimp" was pretty much restricted to one species: *Cragon vulgaris,* or the common European shrimp. This commercially fished shrimp is about three inches long, is gray to dark brown, has a smooth rostrum, and has claws, rather than pincers, on its first pair of legs. With the opening of the New World and expanded fishing waters, new species were eventually identified and harvested for market. Although many closely resemble each other, some have markedly distinct features. Gill structure and egg development, for instance, are two features that scientists now use to classify these shrimps into different suborders. Because of the large variety of species, many are often sold simply as "shrimps," with no other accompanying designation, except perhaps a local or a regional name. And whether it is the widely harvested white shrimp, the brown shrimp, or the common European shrimp, when they reach the market, they are often relabeled, sometimes as "prawns."

Like "shrimp," for hundreds of years the term "prawn" had been primarily used to identify a single species of crustacean found in European waters: *Palaemon serratus*. It was sold alongside the common European shrimp and could be distinguished by its serrated rostrum (a pointed extension of the back above the head), a slightly smaller body, and pincers on the first pair of legs. But as other prawnlike crustaceans were discovered, the term "prawn" soon came to cover quite a variety of forms. In fact, scientists working to classify shrimps and prawns were eventually compelled to form several new genera, such as *Alpheus, Farfantepenaeus,* and *Penaeus*.

But using systematic criteria for classification, rather than their common names, made for quite a mix of "shrimps" and "prawns." In *Penaeus,* the banana prawn, the Indian prawn, and the green tiger prawn sit next to the brown shrimp, the Kuruma shrimp, and the white shrimp. For these, the terms "shrimp" and "prawn" could easily be exchanged; they do not imply any collective difference. Yet not all prawns are treated so lightly. *Palaemonetus vulgaris,* the common prawn, is always considered a prawn, never a "shrimp."

To the British, who take their seafood quite seriously, "prawn" is a much more meaningful term than it is to North Americans. Loosely speaking, they use it for many crustaceans that Americans call shrimps. In a more pedantic vein, many Britons feel that a prawn is a prawn and a shrimp is a shrimp—although the derivation of "prawn" is unknown and its true meaning long lost. The British contend that "pink shrimp" is a misnomer—the crustacean is really a prawn. The distinction, they say, is in the rostrum: those of prawns are pointed and extend over and in front of the eyes, whereas the rostrums of shrimps do not. Also, the first abdominal segment overlaps the second segment, whereas on shrimps the second segment overlaps the first. The British also point out that the Dublin Bay prawn is, in fact, a marine crayfish (to be accurate, it is the Norway lobster). North Americans, while more apt to use "prawn" indiscriminately, do use it instead of "shrimp" for some species, particularly the commercially harvested *Cragon franciscorum* of the Pacific Northwest.

"Krill" was originally a Norwegian word used to identify any kind of buglike creature. Today it is commonly applied to several sea creatures whose only shared attribute is serving as a meal for whales. Depending on the type of whale and its feeding locale, "krill" encompasses mysid shrimps, red crabs, and small fish. But most often, "krill" refers to the favored food of the baleen whale: the crustacean *Euphausia superba*.

Euphausians, a favorite of birds and about 32 species of fish, are so abundant that it has been estimated that in weight, they make up more than half the zooplankton in the oceans. Swimming together, they form shoals that range in size from a few feet across to those of truly colossal proportions. One swarm was estimated to reach a depth of 650 feet and to cover an area the size of Minneapolis, Minnesota (59 square miles). While the species *E. superba* of the antarctic waters is the best known, close to a hundred others complete the genus, all of which somewhat resemble the shrimps of Decapoda.

Several notable differences separate the Euphausian krill, and in particular the two-inch-long *E. superba* species, from shrimps and prawns, but most notably, it is its exclusion from Decapoda. Placed in the order Euphausiacea, this krill has more than five pairs of thoracic legs—usually six. These are all used for feeding, with another five oar-shaped pairs on the abdomen, called swimmerets, which provide propulsion. And because the krill uses its forward legs for feeding, it has not developed maxillipeds: auxiliary mouthparts used by shrimps. Also, the carapace on krill, covering its thorax, does not extend all the way down the sides, leaving its lateral gills, unlike those of the shrimp, uncovered. Finally, whereas the tail of a shrimp is somewhat plain, the krill's tail terminates in a covering of bristles.

Shoals of *E. superba* krill can be a remarkable sight. Because their bodies are translucent, with dark red to rust-colored blotches, when a shoal approaches the surface during the day, the sea takes on a blood-red hue. More spectacular, however, are these shoals when seen at night. The sea will look ablaze in blue-green light, given off by these tiny crustaceans massed together in the millions, an effect produced by the over two dozen light organs on each krill body.

Skate • Ray • Shark

Skates and rays, like their close relative the shark, are Chondrichthyes: fish with cartilage skeletons rather than bone. Although not dangerously aggressive like some sharks, a few skates and rays have features capable of inflicting injury and even death. They differ most noticeably from sharks in possessing flattened bodies (two exceptions are noted further on), with wide pectoral fins attached to the sides of their heads. Somewhat resembling underwater bats, they are occasionally referred to as batoids. Also, unlike the shark, whose gills lie along the side of its body, the gills of skates and rays form on the underside of their heads; and their eyeballs are fixed to the upper rims of the orbits, not free, as in sharks. Furthermore, the tail fins of most batoids are dramatically reduced, if not absent; so, whereas the shark keeps its pectoral fin still and propels itself with its tail, all skates and most rays swim mainly by either flapping or undulating their pectoral fins.

It may seem obvious that sharks and batoids, so different in appearance, would never be confused with each other—and, usually, they are not. However, the skate and ray superorder Batoidea also includes the sawfish (order Prisiformes) and the guitarfish (order Rajiformes). With substantial tail fins and more cylindrical bodies, the two could easily be mistaken for sharks. Yet in those defining anatomical features just cited, they are true rays. Then there is the angle shark, or monkfish. At first sight, this true shark resembles skates even more than do guitarfish, but its gill slits are on its side and the pectoral fins are not attached to its head.

The eye covering and the respiration of skates and rays also differ from those of most sharks. Many sharks protect their eyes during feeding by movable eyelids called nictitating membranes. Skates and rays, whose eyes are on top of their heads (eagle rays excepted) and away from their mouths, have no need for such protection; consequently, nictitating membranes never evolved or were subsequently lost. Most batoids live close to the sea bottom, often resting on or burrowing under the sand. To avoid inhaling sand, they breathe in through openings behind each eye called spiracles (present on some sharks as well) and exhale through their gill slits. This breathing technique is also used when swimming, although surface-dwelling manta rays will take in water through their mouths like sharks. Only the few bottom-dwelling sharks find it necessary to use their spiracles for respiration.

Although the terms "ray" and "skate" are restricted to specific species, this has not always been the case. Not too long ago in the United States (and now in Great Britain), some skates, like the thornback, were also called rays. This particular use of "ray" and "skate" possibly arose from a view expressed in 1753, and cited in the Oxford English Dictionary. In the grouping of "rays," "the smooth are what we call skates and flairs; the prickly we call thornbacks."[3] Earlier writings confirm this notion: skates were smooth bodied and rays rough skinned. Yet this was before modern-day classification took other anatomical features into consideration. Today, both skates and rays may be either smooth or rough skinned. (One note on the use of "ray": when one speaks in general terms, "ray" is often used as shorthand to cover all members of the order. Here, "ray" is used only for specific species.)

"Skate" was originally an Old Norse term that goes back more than 800 years, which was only applied to the common skate, *Raja batis*. With the exception of the guitarfish, "skate" is now used to identify those exclusively of the order Rajiformes. The other batoids belong to an unsettled number of orders, whose membership varies, depending upon the authority. Ignoring the sawfish, the balance of batoids currently makes up two orders: Torpediniformes and Myliobatiformes.

Most features used to identify the skates are mainly variations of ray characteristics. The skate's tail is not as heavy as the electric ray's or as slender as that of the stingray, the eagle ray, or the manta. The skate's pelvic fins are deeply notched. And like the electric ray, many, if not all, skates carry electricity-producing organs. But whereas the electric ray's organs lie within its pectoral fins and can deliver up to 200 volts at 50 amps, skates carry theirs in their tails and only manage about 4 volts.

Despite their many shared features, one unique characteristic does set the skates apart from the rays: their type of birth. All rays are born alive through a process called ovoviviparous birth, in which the young hatch from eggs while still inside the mother. Skates emerge from their eggs only after the mother has laid these on the sea floor, a form of birth termed oviparous. These eggs, occasionally found on beaches, are encased in a leathery pouch that has thin tendrils projecting from each corner, designed to attach to objects. Sometimes called "mermaids' purses," "sailors' purses," or "sea purses," they will hold the skate embryo for up to fourteen months before hatching.

As a group, the electric ray, the stingray, the eagle ray, and the manta are considered to be the "true" rays. Yet no specific feature can be said to be typical of them all, unless it is their type of birth, and even this is seldom given much weight. The group, in fact, consists of families that are just as individually distinct as the skate family is. The following features are unique to each family.

- Electric ray: With an almost-circular body shape, its electric organ, accounting for one-third of its weight, can electrocute large fish.

- Stingray: Bearing a whiplike tail armed with one or two poisonous spines near its base, it is capable of killing humans.

- Eagle ray: With an even longer tail, it also has a distinct head structure, and its eyes and spiracles are located on the side, rather than on top.

- Manta: Up to 22 feet wide, this largest of all Rajiformes carries projecting head lobes, called cephalic fins, on each side of its wide, forward-facing mouth.

Considering the diversity within the order and the ill-conceived distinction once made between skates and rays, the differences between the two forms appear no more significant than those differences that separate the individual families themselves.

Squid ● Octopus ● Cuttlefish

Imagine that you are out at sea, fishing from a small boat, when you spot what appears to be wreckage in the water a few feet away. As you attempt to draw it near with a boat hook, it suddenly breaks the surface and emerges as a huge head with enormous eyes. Lunging toward you, its beak viciously attacks the boat's gunwale. Twenty-foot tentacles lash out, and, wrapping itself around the craft, the creature starts to pull you under. Quickly, you grab a hatchet and desperately begin hacking away at the monster. Slowly, it releases its grip and slides back into the sea, minus a couple of arms.

Farfetched or not, this scenario recounts a true incident. It happened to 12-year-old Tom Piccot, his father, and another fisherman in October of 1873 off the coast of Newfoundland. The beast was a giant squid, a true monster of the sea, but not of record size. One of the largest squid ever encountered was estimated to weigh close to 8,500 pounds, had a body 20 feet long, and sported 35-foot tentacles. Its eyes measured 9 inches across.

As for true monsters, however, the most incredible creature to come from the ocean is thought to have been a giant octopus that washed ashore at St. Augustine, Florida, in 1897. This animal, whose arms were estimated to be between 75 and 90 feet long, may have been able to anchor itself with one arm and to reach out 150 feet. But an octopus—if that is what it really was—approaching this size would be a freak of nature; most are far smaller, which leaves the title of "tentacled monster of the sea" to the giant squid of *Architeuthis* spp. This relative of the octopus is, without doubt, the largest invertebrate animal on earth and has been the genesis of countless legends of sea monsters—some fanciful, others rooted in amazingly true encounters. The dreaded Kraken of Norse legend that attacked sailing ships was almost certainly one of these giants. Fortunately, most other squid are much smaller, usually falling in size somewhere between the fire squid of only a few inches and the 2- to 3-foot-long common American squid.

Octopuses, squid, and cuttlefish belong to the mollusk class Cephalopoda, a name that refers to a well-developed head attached to grappling arms or tentacles. But not all are equally endowed; the squid and the cut-tlefish each have ten arms and an internal supporting shell, whereas the octopus has only eight arms and no shell. All propel themselves by taking in water and forcefully expelling it through maneuverable jets.

Although the use of the term "arm" is more common than "tenta-cle," both are used interchangeably, particularly for the octopus, whose appendages are identical. The cuttlefish, as well as every squid but one, also has eight extremities that are alike, plus two exceptionally long limbs, terminating in sucker-studded paddles that are used for grabbing prey. These are properly called tentacles and the other eight called arms. The one exception among the squid is a recently discovered species of the Magnapinnidae family, whose 23-foot, skinny, ropelike arms are all alike. These arms are not thought to grasp prey but rather to ensnare it.

The body of the octopus, lacking an internal shell, looks much like a partially inflated sack. Although this is sometimes thought to be the head, the head proper, with its large eyes and sharp beak, in fact sits somewhat between the body and its arms. Renowned for its powerful suckers, a good size octopus of 5 feet is reported to have a grasping force of over a quarter ton. Yet these smooth suction cups are no match for those of the squid or the cuttlefish. The suckers of these two are reinforced around their rims by horny rings, which in a number of species also carry sharp, toothlike projections to enhance adhesion. And to the horror of their prey, in some deep-sea squid these "teeth" have evolved into sharp claws, a notable feature of the vampire squid, whose suckers have been replaced by pronged spikes.

Octopuses are solitary creatures. Rarely socializing except to mate, they spend much of their time hidden in dark caves or resting under ledges until hunger compels them to hunt up a meal of crustaceans or shellfish. Such a sedentary lifestyle is in direct contrast to that of the squid. Squid, mainly open-sea creatures, assemble in huge numbers to feed on animal life that ranges from plankton to tuna—and each other. In ferocious attacks, they will often continue decimating a school of fish long after they are sated. Those living near the surface will even propel themselves out of the water in pursuit of flying fish. Attaining speeds of up to 30 miles per hour, they have reportedly been seen to leave the water and sail almost 200 feet. Sailors even report finding them on ship decks 15 feet above the water. Underwater, the only prey that will elude the larger squid are a few very fast game fish.

Cuttlefish are agile swimmers but not nearly as powerful as squid or as voracious. As solitary predators, they glide over shallow sea bottoms in search of smaller food, such as crabs and shrimps. Spotting a meal, the cuttlefish will shoot out its two long tentacles—often kept in pockets located below its eyes—seize its prey, and quickly bite the prey with its beak, injecting a paralyzing poison. To rest and digest the meal, it will bury itself under a thin covering of sand. Cuttlefish are not found along any American coastlines, preferring instead the warmer coastal waters of the Mediterranean and the Indo-Pacific seas.

The cuttlefish, ranging in length from 1½ inches to 5½ feet, is on average a smaller creature than the squid. Most are only about 2 feet long, and where the squid is rocket shaped, with a circular cross-section, the cuttlefish is wider, forming a more flattened profile. This makes a cuttlefish a more massive creature than a squid of equal length. The width of the cuttlefish is also enhanced by its undulating, ribbonlike fin, which on most species extends unbroken around the body. The fins of the squid, resembling two triangles, are attached along its sides, usually toward the rear.

Probably the most familiar feature of the cuttlefish is its porous skeleton, called a cuttlebone. This slightly arched, badge-shaped shell is commonly used in birdcages as a source of calcium. More important—to the cuttlefish, anyway—is its twofold function as structural body support and a buoyancy regulator. Whereas the squid keeps itself balanced by constantly moving its fins and using its jets, the cuttlefish maintains a reserve of fluid and gas (97 percent nitrogen) within the bone, whose ratio it varies to maintain balance. The smaller, pen-shaped bone of the squid has no such function; it is used solely for support.

Squid, cuttlefish, and octopuses all have the ability to change color by subtle muscle control. This is done to camouflage themselves by matching their backgrounds or to display their moods—fear, anger, or contentment. Squid, for example, may slowly go from a pale hue to blood red during a feeding frenzy. For the cuttlefish, this change in color or pattern occurs at incredible speed—in a fraction of a second. No other animal is as quick or as versatile. Besides simple color changes, the cuttlefish will display a zebra striping just before mating or, when in danger, will go pale and produce two "eyes" on its back to frighten predators away.

Not to be outdone in bodily display, some squid light up. About a third of all species have highly developed light organs on their bodies that act as lenses, reflectors, color screens, and shutters. These glowing structures are unsurpassed in the animal kingdom. One species, the fire squid from the Indian Ocean, has 22 of these organs, made up of 10 structural combinations. Unfortunately for the squid fan, most of these bioluminescence squid are small and rarely seen deep-sea dwellers.

5

AMPHIBIANS AND REPTILES

Amphibian ● Reptile

Interestingly, frogs and salamanders are part of a somewhat new group of animals not quite two hundred years old. This is not to say that they suddenly sprang up, croaking and scurrying around, in the early nineteenth century—their ancestry goes back millions of years—but rather that they were previously considered to be reptiles. Lumped together with snakes, lizards, and turtles, frogs and their kin were acknowledged to be different creatures (called batrachians), but no more so than snakes were different from crocodiles. Altogether, these animals made up a large and very diverse group. Initially, the reproductive organs of the batrachians set them apart from the other reptiles; however, when science seriously set out to classify all life, it became apparent that the reptile class had harbored two separately evolved forms. It was decided that the two could best be distinguished by examining their evolutionary histories, focusing primarily on the retention of different primitive characteristics. Scholars began to see a logic in distinguishing between the batrachian reptiles, whose lives included both aquatic and terrestrial stages, which they called amphibians, and those that did not, which retained the name "reptile." They coined "amphibian" from the Greek *amphibios,* meaning, "living a double life." The earlier, all-encompassing "reptile," meaning "crawler," appropriately referred to the common mode of locomotion used by both forms. Today, the class Reptilia consists of the crocodilians, the snakes, the turtles, the lizards, the little-known amphisbaenians, and the lone tuatara—a lizardlike reptile that is fighting extinction on a few islands off the coast of New Zealand. The more recent class, Amphibia, includes the salamanders, the frogs and the toads, and the wormlike caecilians. In the context of taxonomic history, it could be said that amphibians arose from reptiles; however, as a process of evolution, it is the opposite that truly reflects the facts.

Closely examined fossils dating back some 350 million years show a definite evolutionary connection between ancient fish and primitive amphibians. These early amphibians, named Labyrinthodontia because of their folded teeth, were fish-eating, four-limbed, air-breathing creatures that spent varying amounts of time on land. Considered to be evolved descendants of fish, the amphibians retaining scales appear to have been primarily aquatic species, whereas those that developed smooth skin may have been completely terrestrial. Eventually, these Paleozoic amphibians formed two main groupings, which anatomists distinguish primarily by vertebrae structure. Those now classified as Temnospondyli evolved to be our present-day amphibians. The others, the Anthracosauria, eventually became reptiles—some of which would ultimately lead to human evolution. This split within the amphibians occurred some 300 million years ago, although the ages at which various branchings and definite forms emerged are unclear. It is certain, however, that by the Permian period (275 million years ago), reptiles had arrived and were diversifying into forms that eventually became turtles, snakes, lizards, crocodiles, and dinosaurs. At the same time, of course, the amphibians (those of Temnospondyli) were also evolving toward their present forms.

But evolved as today's amphibians may be, they still retain characteristics derived from their aquatic ancestors. Like the skin of fish, their skin, if not slimy, is almost always thin and moist; with few exceptions, they must reproduce in water; and their eggs lack amniotic membranes enclosing the embryos. These features still serve amphibians well, yet in their effort to become land animals, the ancient protoamphibians had to make two important adaptations for survival. Land travel required a new means of locomotion, so fins eventually evolved into legs; second, the gills gave way to lungs, and the scaled skin of fish grew naked, both of which were needed to absorb atmospheric oxygen. This accomplished, amphibians came into their own. Yet their thin, permeable skin and their method of reproduction still kept them tied to aquatic or, at the least, very moist locations. Water was necessary to keep their skin moist and functioning and to keep their eggs, lacking a protective covering, from drying out. Constrained by this dependence on water, amphibians developed various aquatic and semiaquatic lifestyles.

Today, most live typically "amphibian" lives, having an aquatic larval stage (such as tadpoles) and a terrestrial adult stage. Some, however, never leave the water, whereas a few others live their entire lives on land (see

"Salamander • Newt"). These basic amphibian requirements continue to keep most of them from gaining an effective foothold in the world's more arid regions and grasslands. Once pioneers of terrestrial habitation and the precursors of all land animals, amphibians remain generally small, secretive, and vulnerable creatures who now play a comparatively minor role on Earth, although the number of species does exceed that of mammals.

In contrast, the ecological niche filled by today's reptiles is much more substantial. Whereas crocodilians and turtles generally stay close to aquatic environments, preferring wetlands or more open waters, many snakes and lizards adapted to deserts, plains, and rocky terrain—areas hostile to most amphibians. Reptiles not only successfully adjusted to these new environments, but they often became the dominant animal group. They achieved this position by dissolving most of the remaining ties that their amphibian ancestors (those of Anthracosauria) had with fish, as well as by modifying a number of unique amphibian features.

Reptiles took their biggest evolutionary step forward when their eggs developed hard shells that contained nutrients and protective membranes, which allowed them to be laid on dry land without risk of drying out (some reptiles do bear live young, rather than lay eggs). Also, fertilization became internal, a more efficient and reliable technique than the common amphibian method of depositing sperm on previously lain eggs. Amphibian respiration by means of gills and oxygen-absorbing skin was eliminated, and the amphibian process of gulping air into the lungs was replaced in reptiles by negative respiration. This method of breathing, using the respiratory tract and the lungs, works by enlarging the chest cavity, which lowers the air pressure around the lungs and, in turn, compels the lungs to suck in air. With no need for respirating skin, reptiles developed a thick, dry, waterproof skin, which not only kept moisture in but also protected them from the environment. They replaced the toxic skin glands of amphibians, which afford protection from predators, with horny scales in the form of crests, spines, plates, and, in the turtle, a shell that usually employs bony components. As a defense, reptiles also developed claws (the leatherback sea turtle being an exception) and, in some species, jaws that deliver poison.

Whereas these fundamental changes continue to characterize reptiles, the following features, contrasted with those of amphibians, typify other changes that evolved and have become characteristic.

- Reptiles developed stronger limbs and body support.
- All reptiles have the same finger bone formula (2–3–4–5–3), whereas that of the amphibians varies considerably.

- Reptiles have rib cages, whereas amphibians do not.
- Reptile skeletons contain more bone material and less cartilage than do amphibian skeletons.
- Scaled reptiles shed their skin far less frequently than do amphibians.
- Reptiles have more efficient kidney and circulatory systems.
- The nervous system of a reptile, with 12 cranial nerves, is more advanced than that of an amphibian, which has only 10 cranial nerves.
- Almost every male reptile has a copulatory organ, a device extremely rare in amphibians (the tailed frog is one exception).
- The eggs of reptiles have shells, whereas those of amphibians are covered in jelly.
- Reptiles are better able to regulate their body temperatures by changing their behavior or location.
- A few reptiles (some skinks) develop placentas; amphibians do not.
- The reptile skull is deeper and narrower and has a notch cut in the back called the otic arch, whereas the amphibian skull is short and broad and lacks an otic arch.

The evolution of today's reptiles from the anthracosaurian amphibians was certainly a significant one, but it wasn't the only change taking place. Three other distinct branches also developed. One led to a split that now distinguishes today's turtles from the other reptiles, one led to the development of birds, and the most significant of all led to the evolution of modern mammals.

Alligator ● Crocodile ● Caiman

Today's crocodilians—the crocodiles, the alligators, the caimans, and the gharial—represent an ancient branch of reptiles that began its evolution alongside the dinosaurs some 200 million years ago. Surprisingly, crocodilians and birds descended from an extinct reptile-like group of animals distinct from those that gave rise to the lizards. Although distantly related to birds, crocodilians successfully adapted to their environment by developing characteristics similar to those of lizards. In fact, this similarity is reflected in the names "alligator" and "crocodile." Spanish explorers who discovered the alligator along the Gulf Coast of North America named it *el lagarto,* after the Mediterranean wall lizard *lagarto.* Through years of corruption, *el lagarto* eventually became "alligator." Likewise, "crocodile" is said to have evolved from an Ionian word for a small native lizard called *krokodeilos,* meaning "pebble worm."

Although some authorities classify crocodilians as a family embracing three subfamilies, it is more often considered an order, Crocodylia. This order consists of the following three families, each of which took separate evolutionary paths some 65 million years ago: Crocodylidae, containing all the crocodiles; Alligatoridae, composed of alligators and caimans; and Gavialidae, with one lone species, the gharial (also called a gavial and a garial).

The gharial is a slender-mouthed reptile that lives in the northern part of the Indian subcontinent. Alligators, with an only slightly broader distribution, reside in just two parts of the world: the lower leg of China's Yangtze River, home of the small Chinese alligator; and the coastal areas of the southeastern United States, where the American alligator dwells. Caimans live in Mexico, as well as in Central and South America, often inhabiting inland wetlands and waterways. The remaining family, the crocodiles, has a far greater distribution, which includes Central and South America, Florida, Africa, India, Indonesia, and Australia.

Although scientists usually restrict "caiman" (derived from the Carib Indian word *acayoman*) to specific Alligatoridaes, Latin American natives still retain its older, more inclusive meaning, which embraces crocodiles and, in some cases, lizards. Likewise, natives of West African countries, to whom "caiman" was introduced by early French explorers, continue to use the term for local crocodiles. "Alligator," a word best limited to true alligators, has also been somewhat misappropriated. In some of Africa's more eastern countries, natives have adopted "alligator" to identify the Nile crocodile, and in Australia it is occasionally applied to those crocodiles living in estuaries.

With the exception of pronounced variations in skull features, particularly the snout, the general appearance of all crocodilians is very similar. Yet even nominal differences in body characteristics point to a clearly divided evolution, prompting the creation of three distinct families.

Unquestionably, the feature most often cited in distinguishing the alligator from the crocodile is the enlarged fourth tooth on each side of its lower jaw. In the alligator, this tooth, like its others, fits into a socket in the upper jaw and therefore does not show when the mouth is closed. On a crocodile, the snout is notched at this point, which keeps its fourth tooth from being enclosed by the upper jaw and leaves it exposed when the mouth is closed. Another noticeable difference is the general shape of their jaws. The alligator has a relatively shorter and broader snout than does a crocodile. The tip of the alligator's snout is about half as wide as its base, whereas the crocodile's tip is only about a quarter as wide. Also, the alligator's upper jaw is wider than its lower jaw, which permits the upper teeth to overlap and hide its lower teeth, whereas the upper and the lower jaws of crocodiles are about the same width, so the lower teeth fit outside the upper jaw, accommodating a pattern of alternating upper and lower teeth. Another feature of the head, not at all apparent but very helpful in distinguishing crocodilian fossils, is the presence of a unique spinelike protuberance on the side of the crocodile skull.

Both alligators and crocodiles have salt glands on their tongues, glands designed to excrete excess salt water from their bodies. However, whereas these glands are highly functional in crocodiles, permitting them to tolerate saltwater environments, in alligators and caimans these glands have all but shut down, forcing them to keep to freshwater areas whenever possible.

One of the most important features of the alligator—to humans, anyway—has been its belly skin. As a source of leather for making shoes, purses, belts, and so on, alligators were so relentlessly hunted between 1880 and 1933 that in Louisiana alone, an estimated 3.5 million were slaughtered. To help meet the demand for reptile leather, people also hunted the crocodiles of Central and South America; however, the number killed never approached that of alligators. Aside from being geographically less accessible (except for the American crocodile), the main reason fewer crocodiles were killed was their less-desirable skin. Unlike alligators, most crocodiles have belly skin containing bony deposits called osteoderms. Although these hardened plates help protect the belly, they greatly reduce the skin's value—in Australia, the skins of the young are still harvested for the fashion accessory trade. The skin of crocodiles also has unique sensor pits connected to nerve fibers, a feature thought to assist their other senses, which become easily foiled by murky waters. In alligators, these tiny black sensory pits occur only around the upper and lower jaws.

On average, the alligator is a stockier-looking reptile than the crocodile, an appearance fostered by its larger girth, hefty snout, and broader tail. In all, the crocodile is a bit sleeker and more agile looking and actually can move a bit more quickly (up to 10½ mph) than the alligator. Although these general features well describe both alligator species, they are less applicable to the alligators' close relatives, the caimans.

Except for the black caiman, which may grow to 15 feet in length, the other adult caiman species average only 6 feet, somewhat shorter than the American alligator's length of 8 to 12 feet. Not as hefty as the alligator, the caiman moves with the agility of a crocodile and may easily be mistaken for an adolescent croc. Like that of the crocodile, the belly skin of the caiman contains significant osteoderms, yet the caiman remains more closely allied to the alligators (like alligators', the caiman's upper teeth overlap and the lower fourth tooth is hidden when the mouth is closed). Close kinship notwithstanding, the three genera of caimans are set apart from the alligator genus by several features. Most significantly, they have no bony septums dividing their nostril cavities as do alligators. More apparent are the differences in the scuts (raised plates) on their backs. The eight enlarged plates covering the back of the caiman neck lock together more tightly than do the American alligator's six plates or the Chinese alligator's four. Also, the caiman's snout tends to be a bit narrower than that of the alligator, and its tail is generally much shorter.

Behavior among alligators and crocodiles does not differ greatly, but a few traits are notable. The rump of a walking crocodile is carried higher than its shoulders, whereas an alligator's rump and shoulders are kept fairly level. But an alligator need not be seen to be identified. The adult male sometimes announces its presence with a loud roar or a bellowing, a noise that caimans also make but in lesser decibels. If a crocodile does roar, it is not very loud. Usually, it will utter more of a growl or a rumbling sound.

As for habitat, alligators prefer fresh or brackish waters, areas with enough vegetation to construct a large nest. These nests, commonly reaching 3 feet high and 4 to 7 feet wide, are designed to incubate anywhere from 20 to 70 eggs. In contrast, most crocodiles live in saltwater environs where vegetation is not as abundant, but sand or mud is. The mother croc digs a hole in the ground, where she deposits up to 70 eggs, which she then covers with earth and whatever vegetation may be at hand. Both alligator and crocodile mothers closely guard their eggs, but unlike the alligator, who is an attentive parent, making sure that nothing disturbs her brood after birth, crocodiles often leave their young to fend for themselves.

Frog ● Toad

Of all the amphibians in the world, frogs and toads are by far the most common—they comprise 90 percent of all known species. Not only do they delight children with their cautious antics, but these tailless creatures serve as valuable subjects in biology classes and research laboratories. In just as serious but less realistic a vein, they have also been the objects of reverence and fear. The Maya treasured the frog as an attendant to their god Chac; medieval Europeans came to associate frogs, and particularly toads, with the evil concoctions and conjurings of witches. But no matter how they may be perceived, when it comes to distinguishing frogs from toads, their true identities remain clouded.

Originally, "frog" and "toad" were labels for two specific forms of amphibians, best represented by the common European frog *(Rana temporaria)* and the European common toad *(Bufo bufo)*. As other species were recognized, they came to be called either "frog" or "toad," depending on how closely they resembled these archetypes. If they tended to be slim, long-legged, and moist-skinned, and they jumped, they were labeled frogs. If they were fatter, were short-legged, hopped, and had dry warty skin, people called them toads. Naturalists, content with such superficial observations, made no attempt to establish a more detailed distinction between the two for quite some time. However, when scientists did begin sorting out the various species and classifying them into families, all sorts of problems arose.

For a time scholars had difficulty in agreeing on a single classification scheme for the frog and toad order they call Anura (meaning "tailless") or Salientia (meaning "to jump"). Eventually, however, it became apparent that skeletal features, particularly details of the pectoral and pelvic girdle (bones supporting the limbs), and the shape of the vertebrae, would be the best criteria for establishing a taxonomy—since then, various authorities have adopted other criteria. Scientists, looking closely at the morphology, anatomy, and evolution of various species, discovered that while certain frogs and toads had structures suggesting close kinship, their outward appearances were quite dissimilar. And whereas other anurans were superficially very similar, some had, surprisingly, evolved independently of each other. Moreover, none of these criteria unified those anurans popularly called "frogs" into a group distinct from those known as "toads." This made it quite clear that in sorting out various species, little if any weight could be given to the old "frog" and "toad" labels.

With the new guidelines for Anura taxonomy in force, some "frog" and "toad" species ended up as close relatives in the same family. For example, Pelobatidae, the family of spadefoot toads, also contains the parsely frog; likewise, the midwife toad belongs to the same family as the painted frog. Such random occurrences of "frog" and "toad" throughout the order ensured that the two terms could be nothing more than vernacular labels. Today, of the 22 families of Anura, 17 are commonly designated "frog" families and 5 are "toad" families (the exact number of families does vary among authorities). In practice, Ranidae is commonly designated the *true frog family* and Bufonidae, the *true toad family*, but neither label is scientifically meaningful. The other frog and toad families have more descriptive names, such as the *tree* frog and the *burrowing* toad families. Yet some authorities avoid calling any anuran a "toad" except those of Bufonidae and often consider "frog" an appropriate label for all tailless amphibians. Others feel that "frog" should be used only for those of the genus *Rana*. And then there are those who sometimes use the two terms interchangeably for members of the same genus, blurring any perceived distinction between the two labels.

Aside from the few families containing a mix of so-named frogs and toads, most families consist of either all "frogs" or all "toads." Ranidae, one of the largest frog families, has almost 1,000 species; it contains the most widely known species, including the bullfrog, the leopard frog, the green frog, and the common European frog. It is these frogs and the others of their family that reinforce our popular notion of the "typical" frog. For toads, the quintessential family is Bufonidae, with a little over 400 species. Here one will find the American toad, the Rocky Mountain toad, and the European common toad, all having "typical" toad features.

The other 20 families of Anura contain less-familiar species, although the families of spadefoot toads, tree frogs, and narrow-mouthed frogs are well represented in the United States. Almost all other species live outside North America, some in families whose range is very limited, like that of the Seychelles frog family, whose habitat consists of two islands with a combined area of about 65 square miles.

Of the familiar Ranidae frogs and Bufonidae toads living among our North American ponds and streams, the most abundant and typical belong to the genera *Rana* (frogs) and *Bufo* (the only genus of Bufonidae toads in North America). Species of both exhibit all the contrasting frog and toad characteristics established years ago as so typical: slim-fat, long legged-short legged, moist-dry, smooth-warty, jumping-hopping. Yet as creatures from not only separate genera but also separate families, they differ in less obvious but more fundamental ways.

In general, they exhibit only minor variations in habitat and lifestyles: toads tolerate a greater loss of body water and have adapted to a more terrestrial way of life than have frogs. However, taxonomically, the most important difference between the frogs of *Rana* and the toads of *Bufo* is the variation in the skeletal support of the front limbs—the pectoral girdle. In these frogs, the right and the left halves of the girdles are fused, whereas in the toads they only overlap; this allows these toads, unlike the frogs, to better expand their chests. Another salient characteristic is that each toad has a so-called Bidders organ, a unique, rudimentary ovary that develops in the males. Other differences, not quite as defining:

- Almost all the frogs have teeth; none of the toads do.

- The parotid glands (skin glands containing poison) are enlarged in toads.

- Frogs typically lay eggs in clusters. Toads, which are more prolific, lay theirs in strings.

- Most frogs have dorsolateral folds—two parallel ridges of skin running down their backs. Toads do not.

- Many frogs have webbed toes, whereas no toad does.

- The skin secretions of toads are more toxic.

- Defensive behavior in toads often consists of lifting the forelimbs and pumping the lungs full of air to appear larger, whereas frogs tend to flee.

Though these features distinguish *Rana* from *Bufo,* they are not necessarily exclusive. Specific frogs and toads of other genera and families share many of these same features.

Finally, a caution. Although most people are aware that handling a toad will not give one warts, all anurans, the toads in particular, do secrete toxins through their skin. Most are not harmful to humans, but if the toxin of some species is transferred by hand to the eyes or the mouth, severe inflammation can occur. So, regardless of an amphibian's benign appearance, it is advisable to wash one's hands after handling any frog or toad.

Newt • Salamander • Eft

Of all the amphibians in the world, only salamanders are regularly mistaken for reptiles. Many so closely resemble lizards that only their lack of scales and clawed toes—features common to all limbed lizards—gives them away. This distinction, however, was never a consideration of the ancient Greeks. They observed that these amphibians, sometimes hiding in logs that were thrown onto a campfire, often escaped by scurrying through the flames, leaving the impression that they were born of fire or at least were fire-proof. Regarded as a special form of lizard, the Greeks called them "salamanders," meaning "fire lizards": a term that has since come to be used for other items associated with fire, such as portable space heaters, ovens, and asbestos.

The term "newt" has an equally peculiar derivation. Originally an Anglo-Saxon word spelled *efete* or *evete*, it was used for those salamanders, particularly of the British Isles, that happened to have a semiaquatic style of life. With the evolution of Middle English (c. 1150–1475), the term became *ewete*, whereupon the amphibian was referred to as "an ewete." Through loose pronunciation, the article "an," carrying the sound of the letter "n," was split, leaving the "a" standing alone and the "n" appended to *ewete*, changing the form of address from "an ewete" to "a newete"—with the last two "e"s eventually being dropped. In another alteration of the old word *efete*, the last two "e"s were also discarded, leaving "eft," a term used today for immature newts in their terrestrial stage of life.

The salamander and newt order, Caudata, is comprised of about 450 species gathered into 9 families. Although salamanders and newts are often referred to separately, newts are actually a form of salamander. The newt family, Salamandridae, also contains several "conventional" salamanders—most notably, the brook salamanders and the fire salamanders. Whereas most authorities recognize 10 to 14 genera of newts, the British often restrict "newt" to those salamanders of *Triturus*. This genus contains the only three species of salamanders living in Great Britain: the smooth, the palmate, and the crested newts. Of North America's two newt genera, *Notophthalmus* and *Taricha*, six species live in the United States, including the prevalent red-spotted newt of the East and the California newt.

Caudata is an incredibly diverse order. Features found in some species but not others include lungs, external gills, gill slits, hind legs, eyes, eyelids, upper jawbones, immobile lower jaws, tongues, and lachrymal bones. Some can voluntarily discard their tails when endangered, some have toxic skin, some are born fully developed, and some spend their entire lives in a larval state.

Salamanders exhibit a variety of lifestyles, including wholly terrestrial, wholly aquatic, and amphibious. Because salamanders have very permeable skin that easily loses moisture, those living their entire lives on land must stay out of the sun and avoid hot and arid regions. During summer days, they seek cool, damp shelter, often under rocks and logs or buried in the earth. Only at night when the air cools do they come out to feed. After breeding, terrestrial salamanders lay their eggs in a variety of places. Some look for nothing more than a damp shelter, such as the base of flower leaves—a favored egg-laying site of the arboreal lungless salamander. Others deposit their eggs in dry ponds during the autumn months, where they remain dormant until winter or spring rains revive them. Those eggs laid on land retain the developing salamanders through their larval stage and hatch only after the young have developed limbs. Most salamanders, however, lay their eggs in ponds, streams, lakes, or other aquatic spots, where they hatch free-swimming larval forms—similar to tadpoles.

For the completely aquatic salamanders, such as the 5-foot-long giant salamander, hot sun and summer heat pose no problem. Never leaving the water, most aquatic salamanders retain the external fringed gills they formed during their larval stage. The Mexican axolotl is a well-known example of this form of arrested development, a condition called neoteny. In the Asiatic salamanders the lungs are reduced in size, if not missing entirely. Their close relative, the North American hellbender, uses gill slits and lungs but also absorbs oxygen from the water through its skin, or it may surface to gulp in air if the water is poorly oxygenated. Almost all aquatic salamanders lay their eggs in water. The olm, which lives in the subterranean waters of caves, is somewhat of an exception. Some of these odd species retain their eggs in their oviducts until the young are almost fully developed.

Newts, which live various stages of their lives in and out of water, are the amphibious members of Caudata. This lifestyle fundamentally sets them apart from all other salamanders. In fact, the inclusion of some of the more typical salamanders in the same family as the newts is a good indication of just how little newts differ anatomically from other salamanders. Yet these small amphibians are not without their own set of unique features. The skin of most salamanders often feels slimy, whereas newts are never more than moist to the touch. When newts live on land, their skin is rough textured, whereas the skin of other salamanders is generally quite smooth. Almost all salamanders have definite vertical indentations along their sides, called costal grooves; on newts, these grooves are indistinct. Common to each member of the newt family is the double row of teeth centered in the upper jaw, which diverges as it runs back toward the throat.

Newts are probably best known for the physical change they undergo after entering water to breed. As with many salamanders, the typical adult newt is terrestrial, living off the land until it comes time to breed. Taking to the water, newts will spend considerable time mating, but unlike other salamanders that return to land immediately afterward, newts generally remain aquatic until late summer. In its terrestrial form, the newt body is designed for land travel and, conversely, not well suited for long periods underwater, so when breeding time approaches, the newt adapts. Its back may develop a finlike ridge of skin, and its toes grow webbing, but the tail changes most radically. Normally somewhat round, it flattens vertically, with the upper and lower edges becoming very thin and broad like a paddle, which then becomes the newt's main means of propulsion. During this aquatic stage, the skin of the newt also changes, going from rough or warty to smooth, and is usually accompanied by changes in color and pattern.

The larvae, hatching from eggs laid on underwater foliage, normally develop into terrestrial adult forms. Some species, however, remain in a juvenile state while on land and mature only after returning to water to breed. These immature newts are called efts, and for some, this terrestrial phase may last up to seven years. Yet a few newt species spend very little time on land, retaining rudimentary gills their entire lives.

Python ● Anaconda ● Boa Constrictor

In their heyday, no self-respecting jungle B-movie or serial would be complete without a dire encounter with one of these creatures. Whether the snake dropped out of a tree or slowly crept up on an unsuspecting victim, the spectacle of the hero or heroine trying to struggle free from the grip of a 15-foot serpent always made for good adventure and drama. Although such scenes served as exciting movie fare, they all grossly exaggerate a more mundane truth: few verifiable accounts exist of any of these snakes successfully overpowering a human, and in those cases found to be true, most involved children. But this does not mean that the snakes are incapable of swallowing a small adult. The African python has been known to gulp down gazelles, impalas, and even a leopard, and an anaconda was once found engrossed in preparing a meal by wrapping itself around a 6-foot caiman.

Contrary to popular belief, these reptiles do not kill by crushing their victims to death. This is confirmed by dissections revealing that swallowed prey rarely had broken bones. Instead of expending energy in squeezing the breath out of its victim, this snake waits until its prey exhales. It then tightens its body around the animal, which does one of two things. It can prevent the animal from expanding its lungs, which in effect suffocates it to death or renders the animal unconscious, or the pressure can collapse the animal's veins and cause circulatory shutdown. In either case, the snake then engulfs one end of its prey (almost always the head) and begins to slowly swallow. This process is aided by multiple rows of long, sharp, rearward-facing teeth that act like ratchets, ensuring that the victim goes in one direction only, into the stomach. Because of its method of capturing prey, it has no use for a set of grappling fangs, so all of its teeth are alike.

As members of the constrictor family Boidae, these three snakes belong to two very similar subfamilies: Boinae, which contains the boa constrictor and the anaconda, and Pythoninae, home to numerous python species. Both the boa constrictor and the anaconda are specific boa species found only in the tropical Americas. Boas of other subfamilies that live in Madagascar, North Africa, Western Asia, and some South Sea islands are lesser-known and generally smaller species than their New World counterparts. In contrast, pythons do not inhabit the Western Hemisphere. The two dozen or so python species of Pythoninae make their homes in central and south Africa, India, Southeast Asia, Indonesia, and Australia. One possible exception is the Mexican dwarf boa *(Loxocemus bicolor)* of North America, also called the Mexican burrowing python, which, in the opinion of some authorities, should be classified with the pythons.

Of all the large constrictors, the anaconda is the most aquatic, spending much of its time in quiet streams, marshes, or swamps, where it awaits the approach of thirsty forest animals. Its close cousin the boa constrictor, however, avoids water and is more likely to be found living in jungle trees, agricultural fields, or savanna land. Pythons, a varied group of species, inhabit all types of locales: streams, brush, rocky outcrops, jungles, and human settlements.

Because of the routine failure to properly identify snakes, almost all the constrictors have at one time or another been popularly called "boa constrictors"—the snakes most often seen as epitomizing constrictor power and danger. But to use the term "boa constrictor" to confer such qualities on the others is a poor choice because the boa is far from being the largest or the most dangerous of the constrictors. In fact, snake handlers generally regard the boa as a fairly harmless reptile. Its average length of around 10 to 12 feet is less than half that of the anaconda and the rock python. Although normally irascible when in the wild, the boa is an easily tamed snake and, consequently, was widely used in sideshows. But a full-grown anaconda, at 25 feet long and close to 200 pounds the largest snake in the world, is far from being as friendly or pliant. This powerful snake reportedly killed a wading native by pulling him underwater. Similar stories tell of large pythons attacking women who were washing clothes and fishing alongside streams.

In considering those pythons whose reputations are on par with that of the boa and the anaconda, three species stand out. The largest, the reticulated python, lives in the forests of Southeast Asia; the Indian or Burmese python, also of Southeast Asia, dwells in forests or along still waters; and the African rock python, of central and southern Africa, lives among the brush and the rocks. Of the three, the Indian python is probably best known to the outside world. Its brightly colored skin and docile nature have made it a favorite with exotic dancers, snake charmers, zoos, and circuses.

As indicated, the subfamilies Boinae and Pythoninae are very similar, both containing constricting snakes of considerable size—as well as species of more moderate lengths. Curiously, each shares one other small but notable feature: claws on both sides of its anal scale. These are vestigial remnants of hind limbs—clues to the evolutionary kinship shared by snakes and lizards.

Anatomically, there is not a great deal of difference between the boa constrictor and the anaconda or among the various python species. Some of Boinae have labial pits (heat-sensitive organs located between the scales bordering their mouths), but in the boa constrictor and the anaconda species, these organs are absent. Pythons, however, do carry such sensors, although their pits are *within* the scales surrounding their mouths. More commonly, anatomical differences between the various species within each family involve size, coloration and patterning, and arrangement of scales.

Probably the most apparent feature separating the pythons from the boa constrictor and the anaconda is their method of reproduction. All pythons are oviparous (egg layers). Depending on the species, they will lay 5 to 100 eggs, which the female incubates by encircling the clutch, a task sometimes lasting over 3 months. All the snakes of Boinae are ovoviviparous (live bearers). Boa constrictors, on average, give birth to 20 to 60 wriggling youngsters, ranging in length from 12 to 20 inches. The anaconda usually produces 20 to 40 larger babies, which average between 2 and 3 feet long.

Anatomically, four features distinguish the two subfamilies. One, involving the skull, is taxonomically significant. This is the presence in pythons of a supraorbital bone on the roof of the skull above the eye sockets, a feature not present in the Boinaes. Second, most python species carry teeth on their premaxillary bones: again, a feature not shared by the boas and their kin. A third distinguishing feature is the two rows of scales on the underside of the python tail, one more row than that of most Boinae species. Finally, the python has only one fully developed lung, whereas the boa has two.

Snake ● Lizard ● Skink

Considering the focus of this book, to say that lizards and skinks are easily confused with snakes probably appears as absurd as saying that we
tend to confuse lions and tigers with monkeys. Any similarities among the
three reptiles are overshadowed by one striking, commonly perceived difference: whereas lizards and skinks have legs, snakes do not. But as obvious as this popular distinction may seem, it is untrue. The little known or
remembered fact is that a number of lizards also lack legs! Many of these
legless reptiles, such as the javelin lizard, the California legless lizard, and
the glass lizard, so closely resemble snakes that more often than not, they
are mistaken for their serpentine brethren. Because of the close evolutionary kinship between lizards and snakes, it would appear that limbed
lizards are simply snakes with legs, or vice versa. Of course, this begs the
question: If there are lizards without legs, why don't we call them snakes?

In reconstructing the evolution of reptiles, it appears that primitive
lizards came on the scene about 180 million years ago. These animals had
scaly skin, produced shelled eggs, lived on land, and had four legs. It is
believed that around 120 million B.C.E., one form of these lizardlike reptiles began spending much of its life underground. These unusual creatures retained their scales and continued to lay shelled eggs, but their legs,
becoming more of a hindrance than an asset, were eventually lost.
Another theory, based on a 97-million-year-old snake fossil with limbs,
suggests that snakes derived from seagoing lizards. Whichever theory is
correct, evidence of a lizard-to-snake transformation—remnants of
ancestral hind limbs and a pelvic girdle—can still be found in a few of
today's snakes. The pythons and boas are prime examples, both bearing
conspicuous "claws" by their anal areas—considerably altered vestigial
limbs handed down from their lizard forebears.

This was not, however, the only change these evolving snakes-to-be
underwent. To protect their eyes, the ineffective eyelids of lizards were
replaced by transparent coverings of skin tissue called spectacles. The
creature also lost its middle ear, eustachian tube, and tympanum,
although it did retain parts of the inner ear. Snakes also lack ear openings,
a condition often said to set them apart from lizards, although a few
lizards are also without ear holes, making the distinction unreliable. Some
snakes, once thought to be deaf, have since been found to hear sounds
within a narrow range of frequencies.

Yet these evolutionary changes have not been exclusive to snakes. As
pointed out, certain lizards also lost their limbs; others, like most of the

geckos, have spectacles, instead of eyelids, covering their eyes; and whereas the earless monitor lizard can detect sounds, it does so without benefit of a tympanum. These adaptations in lizards did not, however, parallel those occurring in the evolution of snakes. They took place much later in lizard development and are considered atypical, most showing up in lizard species whose families have members without such anomalies.

But we don't refrain from calling legless lizards "snakes" simply because of minor differences in ear structure. As they evolved, snakes further distanced themselves from lizards by losing other structures, as well as by developing new ones. One feature that the snake lost yet was retained by the lizard is the pectoral limb-girdle. Although a few lizards no longer have front limbs, they do retain these bony support structures—a legacy from their four-legged ancestors. The mechanism by which snakes focus their eyes also changed. Unlike lizards and the long-nosed tree snake, which focus on objects by changing the shape of their lenses, all other snakes lack retractor muscles connected to their eyeballs and must therefore focus by moving their lenses forward and backward. Snakes also developed upper and lower jaws made of loosely connected bones, allowing each jaw to spread apart to accommodate the swallowing of whole animals. This feature, coupled with the loose suspension of the lower jaw, allows a snake to engulf an animal four to five times larger than the size of its neck. Another change in the snake skull was the loss of the lachrymal bones (bones in front of and below the eye sockets).

A more obvious difference between the two is the single row of broad scales running down the belly of the snake, which aids in forward movement. In only a few species, such as the blind snakes, have these plates failed to develop. Every lizard possesses numerous belly scales, which for those without legs means that movement can be accomplished only by undulating their bodies back and forth—a common but often optional method of locomotion for the snake. Internally, the large number of vertebrae supporting the snake's back, estimated to range from 150 to 450 bones, distinguishes it from the lizard. The snake also differs internally from the lizard by its lack of a urinary bladder: the small amount of urine produced is absorbed by its feces before defecation.

The skinks, sometimes thought of as reptiles distinct from lizards, actually form a prominent lizard family (Scincidae), with about 1,000 species. The most distinguishing features of these small lizards are their long bodies and whiplike tails, which are covered with very shiny and tightly interlaced scales. Sometimes confused with the whiptail lizards, skinks do not have the enlarged belly scales common to the whiptails.

Also, the legs of skinks tend to be small and their toes quite short; in fact, nearly two-thirds of all skink species have reduced limbs. Some, like the sand skinks of Florida, have tiny two-toed rear legs and one-toed front legs. Species elsewhere in the world are completely legless. These nearly and completely legless skinks are noted for their serpentine movement, enabling many of them to swim like fish through loose sand.

Skinks are quick and secretive. Other than darting about during the day in search of food, they spend most of their time hidden or burrowed under rocks, in plant cover, or underground—only a few are arboreal. This somewhat subterranean lifestyle is made possible by certain physical modifications skinks have adopted, particularly those protecting their sense organs. The ears of most skinks have eardrums situated well within the ears and lie behind openings usually reduced in size and completely covered. Nasal openings are likewise often very small and protected. The eyes of skinks tend to be small, and in the earth-burrowing species, their eyes gain protection from transparent plates that replace much of the lower eyelids. Like many other lizards, most live on insects, whereas larger species may also feed on vegetation, small vertebrates, or carrion. Skinks, widely distributed throughout the world, inhabit deserts, woodlands, mountains, and grasslands.

Turtle ● Tortoise ● Terrapin

More ancient than any mammal, these hard-shelled reptiles have survived relatively unchanged for over 200 million years. Numerous fossils from the Triassic period clearly show that they lived just as they do today, in unmistakable aquatic and terrestrial habitats. The most notable differences between these ancestral creatures and today's descendants are the fossil indications of teeth, an inability to withdraw the head into its shell, and the articulating digits of sea turtles. Unlike their enduring form, however, the grouping of specific species as turtles, tortoises, and terrapins has varied considerably over the last 400 years. Even today, these three terms are applied with exasperating inconsistency.

As with many other plants and animals suffering under two or more confusing names, the major problem with "turtle," "tortoise," and "terrapin" is, in part, one of North American terminology versus British terminology. This transatlantic clash in nomenclature even involves the scientific name of their order.

In North America and elsewhere (e.g., Australia), the preferred term for the turtle order is Testudines (Testudinata), whereas Europeans, particularly the British, prefer to call it Chelonia, an older word. Curiously, the British, who first perceived chelonians to be types of tortoises, chose to name the order by using the same root term (chelon, from the Greek *khelone*) that was used to denote the sea turtle family, Chelonidae; whereas North Americans, who look upon Testudines, as a whole, not as tortoises but as turtles, borrowed the order's name from the tortoise family Testudinidae (from *testa,* for "shell"). The origins of these differences and the varied use of "turtle," "tortoise," and "terrapin" have a long and varied history, one that arose around three hundred years ago.

Prior to the seventeenth century, the term "turtle" did not exist. Before then, all hard-shelled reptiles were called tortoises, or in Spanish, *tortugas.* It is speculated that when the British encroached on Spanish-held territories of the Caribbean and came across the various marine tortoises, they adopted the Spanish term *tortuga* for the creatures, which they subsequently altered to "turtle." This revision eventually encompassed all marine turtles but no other forms in the order. So the English came to recognize three principal types of chelonians: marine turtles, which they simply called "turtles"; land tortoises; and water tortoises (some also differentiated between the side-neck and soft-shelled tortoises). Because neither land tortoises nor water tortoises (fresh- and brackish-water species) were native to Great Britain, they were virtually ignored by the British, who were quite content with these two-word terms and in some cases used "tortoise" to embrace all members of the order.

However, British colonizers of the New World, faced with an abundance of new and unfamiliar freshwater Chelonians, eventually found themselves compelled to devise names for these varied species. But to use "water tortoise" for such an overwhelming variety was clearly an unwieldy constraint. Fortunately, "turtle" happened to be not only handy but also succinct, so a name like "snapping turtle" easily won out over a more awkward "snapping water tortoise." However, such alterations never made their way back to England. Thus, for the British, "water tortoise" and its alternative "pond tortoise" remain popular terms, with "turtle" continuing to be reserved only for marine species. (In some cases, the British will use "turtle" when it is part of a freshwater turtle's proper name.) Eventually, the recognition of marine species as turtles, coupled with the discovery of so many freshwater species, led early Americans to regard turtles as the dominant form of chelonians, prompting them to view the entire order as one of turtles. Although all species of Testudines are considered turtles, in North America "tortoise" is generally the preferred term for those belonging to the Testudinidae family.

Without doubt, the tortoises of Testudinidae are unique; they are slow-moving clubfooted reptiles that have adapted to living on land. The tortoise also carries a domed shell into which its head and tail can be hidden from predators. Its heavily scaled limbs, pulled across its front and rear ends, provide additional protection, making its softer body parts almost impossible to reach. All four of their limbs have sharply clawed, short, webless toes, its front legs taking on a flattened shape useful in digging and the hind legs looking more pillar-like, with "elephant feet." Although they are noted for their immense size, particularly the giant tortoises of the Galapagos Islands, the fully-grown speckled tortoise is only four inches long.

Unlike the majority of those in the order, which are primarily carnivores, most tortoises are herbivores. And whereas a few species inhabit forests, where they feed on selected fruit and plant parts, many live in arid regions such as steppes, savannas, and deserts, where their limited choice of food does not allow them to be as picky. Living under such hot and arid conditions has compelled tortoises to adopt one of two methods of survival. Where at least some water is available throughout most of the year, they establish a daily routine of early morning feeding and working, afternoon cooling in the shade or burying themselves in the ground, evening foraging, and then sleeping at night. Regions with seasonal drought do not allow for such a daily routine but instead force tortoises to gorge themselves with food and water during the rainy periods and then bury themselves for the duration of the drought. In areas with extremely long dry periods tortoises can store enough water in their bodies to keep themselves buried alive for months at a time. To retain needed water, the tortoise does not eliminate toxic ammonia from its system by urinary flushings, as do the highly aquatic turtles, or by converting it to the less-toxic urea, a fluid produced by the more terrestrial turtles. Instead, the tortoise changes this nitrogen waste by-product to uric acid, a safe semisolid with very little water content. But not all tortoises live in arid localities. Some tropical rain forest species relish their humid climate and find shallow bodies of water appealing, particularly as a source of food. Here, not only plant but also animal life is sought out, including insects, crustaceans, fish, and frogs.

Yet not only does such tortoise life contrast sharply with the norm; a few turtles have adopted equally different lifestyles from their aquatic kin. The box turtles of North America, which can be found anywhere from the hardwood forests of the East to the grasslands of the Southwest, generally use lakes and ponds only for cooling off. Certain "pond turtles" of

Southeast Asia also live inland, away from any bodies of water, feeding on both forest floor creatures and vegetation. Surprisingly, some populations of these turtles have so completely adapted to terrestrial life, they will drown if placed in water.

With the exception of these unconventional species, most all other turtles live in or around water—their common bond. Yet terrapins are sometimes singled out as if they comprised a third distinct group within the turtle order. The German amphibian scholar Fritz J. Obst suggests as much in his book *Turtles, Tortoises,* and *Terrapins.* Unfortunately, making this distinction is misleading because the terrapin is quite an ordinary turtle.

The word "terrapin" comes from the Algonquin *toarebe* or *turupem,* meaning "little turtle," and is most frequently applied to *Malaclemys terrapin,* the diamondback terrapin. This small to medium-sized (4"–9½") salt marsh turtle was, at one time, an incredibly abundant species and until the 1930s served as a favored gourmet delicacy. Its popularity is said to have developed not because it was superior to other turtle meat, but because it was merely the fad food of its day. This unusual demand for its meat during the Gay 90s and into the first quarter of the twentieth century pushed its selling price up to an incredible $120 per dozen. Today, "terrapin" is still associated with exceptionally good-tasting turtle meat.

As a distinct species, the diamondback terrapin is not all that unique (it belongs to the same family, Emydidae, as do the wood turtle, the box turtles, the map turtles, and a host of others); its most notable feature is simply the pattern of concentric diamonds sculpted on its carapace (upper shell). It is also recognized for the broad, crushing surface of its upper jaw and the absence of stripes on its head and neck. At the time of the diamondback's popularity, at least fourteen other species in other genera were marketed as terrapins—an attempt to cash in on the demand for turtle meat. Accordingly, Raymond Ditmars, an authority on reptiles at the time, defined terrapins as any "hard-shelled, fresh-water, species that are edible and have a recognized market value."[1] This, no doubt, was the reason for the large number of "terrapin" species years ago, and why today only the diamondback is regarded as a terrapin—the others having fallen out of favor and subsequently reverting to their old "turtle" names.

However, the opportunistic British saw "terrapin"—a term they linked to small freshwater chelonians—as a welcome alternative to their clumsy "water tortoise." Although this older term remains in use, "terrapin" is now often used to denote an equally broad array of species. Its

British use is freely applied to species ranging anywhere from the Indian keelback "terrapin" of Southeast Asia to any turtle kept in an aquarium.

And if this were not confusing enough, an entire genus of freshwater turtles is labeled *Terrapene*. These are commonly known as American box turtles, and though belonging to the same family as the diamondback terrapin, none were marketed as a terrapin or ever considered to be a desirable table dish. Living in semiarid to forested regions, their genus name derives not from the Algonquin language but from the Latin *terra,* for "earth."

6

BIRDS

Blackbird ● Grackle ● Crow ● Raven ● Rook

There are blackbirds and then there are *black* birds. Whereas black birds are predominantly black—no surprise—the same cannot be said of all blackbirds. Some "blackbirds," such as the "four and twenty" that were "served before the king," were not even true blackbirds, although they were black. These unfortunate two dozen of nursery rhyme fame were actually black thrushes, which at the time of the rhyme's debut were the only recognized "blackbirds" in Great Britain. Not surprisingly, English colonists in North America, upon seeing the resemblance of indigenous black, robin-size birds to the black thrushes of Europe, began calling these New World birds "blackbirds." Such casual use of Old World names was quite common back then, especially when the resemblance was close. Settlers of the New World also did this with the meadowlark, the oriole, and the buzzard.

Today, blackbirds help make up an entirely new family of exclusively New World birds, Icteridae (ironically, meaning "yellow"), which also includes orioles, meadowlarks, cowbirds, bobolinks, and grackles. The black thrush of old English cuisine has retained its name "European blackbird" but remains unconnected to the New World blackbirds. Although the Icteridae family and its members are loosely called blackbirds—black being the dominant color among its species—many, such as orioles, are far from black. Therefore, this assessment of "blackbirds" focuses only on those species with predominantly black plumage.

Blackbirds are principally flocking and colonizing birds, some gathering into populations more than 100,000 strong. Their diet ranges from seeds and fruits to insects, fish, small amphibians, and even other birds in times of famine. Presumed to have evolved from small finchlike birds, those in the United States and Canada range in length from 7½ to 18 inches. The sexes are dimorphic, meaning there is a distinct difference between the two, particularly in size and color. Males are larger but generally not by more than 10 percent. Males are also much blacker; the females' plumage takes on the more camouflaging hues. The so-named blackbirds of North America that are predominantly black are all short-billed rural species under 12 inches long. Included in this group are the red-winged blackbird, the tricolored blackbird, the yellow-headed blackbird, the rusty blackbird, and the brewer's blackbird. Of these, only the last two are completely black—as are the three species of grackles.

Grackles, which range from North to South America, often frequent urban settings. These larger (11 to 18 inches long) glossy black species are set apart from the other blackbirds not only by their size but also by their long keel-shaped tail and slender bill. And unlike almost all the others, in bright sunlight their plumage will shine with iridescent greens and purples (the brewer's blackbird is somewhat iridescent). Sometimes called jackdaws, grackles are unrelated to the jackdaw of Europe, which is a member of the crow family.

Besides the jackdaw, the crow family (Corvidae) also includes coughs, jays, nutcrackers, pies, ravens, rooks, and, of course, crows. Of the nearly two dozen genera within the family, it is *Corvus* that contains the crows, the ravens, the rooks, and the jackdaws. Whereas "crow" is the traditional name for its genus and even the entire family, in common usage it usually refers only to specific species of the *Corvus* genus—which is how it is assessed here.

Crows are somewhat solitary birds, particularly during the nesting season. And although the flocks they form are typically not nearly as large as those of blackbirds, the American crow has been known to assemble into flocks numbering close to 200,000. The range in diet of crows, like that of blackbirds, is quite wide. Known for their scavenging, often around garbage dumps, they are not above feeding on carrion—also a particular favorite of the raven. As with most blackbirds, not all crows are completely black. Numerous species in North America, Europe, Africa, and Southeast Asia bear various amounts of gray to white plumage on their necks or breasts. Such coloration among crows is consistent in the sexes, as is size—they are not considered dimorphic.

Among the differences that separate the crows (and ravens) from the blackbirds, three are said to be fundamental: the number of primary (outer) wing feathers: blackbirds have nine, crows have ten; the arrangement of tarsal (leg) plates: on crows they overlap, on blackbirds they do not; and the presence of bristles at the base of the bill: blackbirds, unlike crows, have none. If these were the only differences between the two, it would literally take "a bird in the hand" to tell which was which. However, there are several conspicuous features that more easily distinguish the two, the most obvious being size.

Because this discussion focuses primarily on those birds of *Corvus* that are black and live mainly in North America, many species such as the jackdaws and various other multicolored crows are not included in this assessment. (Jackdaws are European and Asian species that superficially resemble blackbirds more than they do the typical crow.)

With the exception of the grackles, crows are much larger than blackbirds. The smallest North American crow, the Mexican crow, is about 14½ inches long, and the largest, the American crow, may reach 21 inches in length. A grackle can be readily distinguished from a crow of equal size by its relatively long tail—a telling feature that holds true in distinguishing all blackbirds from crows. When a crow or a raven is at rest, its wing tips will cross over its tail, a wing position never occurring with the blackbird. Also evident is how the head of the crow and raven rests on an extended neck, whereas the blackbird's head sits closer to its body. Finally, there is the distinctive bill of the crow and the raven, especially those of North America. These New World species all have very husky beaks, unlike the rest of the *Corvus* species throughout the world, which have more slender bills. Their upper jaw can be quite massive and the lower jaw very stout, its bottom edge extending horizontally back to its base. In contrast, the blackbird bill is much smaller and quite pointed, with its bottom edge angling downward from its tip.

Of the 27 *Corvus* species in the world popularly called crows, the four living in North America (the Mexican, the American, the northwestern, and the fish) are all black. Only one of North America's two ravens is completely black, the common raven. The other, the Chihuahuan raven, has snow-white feathers at the base of its neck, which may or may not be conspicuous. And unlike crows, the throat and breast feathers of all ravens taper to a point, creating a wild, shaggy look. The common raven, with a length of 27 inches, is the largest of the perching birds, whereas the Chihuahuan raven is no larger than the American crow. As a large, shaggy bird with a very robust bill—not unlike the jaws of a bolt cutter—and somber color, the raven is a very sinister-looking creature, worthy of its reputation as a bird of ominous portent. The crow, with its sleeker look and smaller bill, is only a somewhat less malevolent omen of folklore. In flight, the two can be easily distinguished by their tails. The raven's tail is distinctly wedge shaped, whereas that of the crow is squared off or a bit rounded.

Crows, unlike ravens, are less wary of civilization and may be found in urban areas. The raven is primarily a wilderness bird, usually keeping to rugged terrain such as coasts, cliffs, canyons, mountains, or deserts. Its call is equally rugged, a somewhat harsh "crrruuuck." The voice of the crow is the more familiar and less irritating "caw" or variations of such a call. Sky borne, crows seldom glide for more than a few seconds before flapping, and when they do glide, their wings are bent upward. When the raven takes to the air, it will often soar on updrafts, occasionally delighting in aerobatics, or it will hover motionless in a strong wind, positioning its wings straight out, and will do very little flapping.

There is just one rook in the world. It is an all-black species whose distribution extends across Europe and Asia. This social bird is well known for nesting in large colonies called, in its honor, rookeries—a term now adopted for similar nesting birds such as penguins. Unlike the crow's and raven's, the rook's bill is unusually pale in color and its face bare of feathers, giving the impression of a high forehead. In size, behavior, voice, and flight, it is closer to the crows than to the ravens, yet its somewhat wedge-shaped tail resembles that of the raven.

Buzzard • Vulture

"Buzzard" is a word that easily evokes a host of grim images. Probably none comes to mind more readily than the old cartoon depiction of a band of hungry birds circling high in the sky above a barren desert, their huge black wings held motionless as each bald head fixes its eyes on a parched figure crawling through the blistering sand below. As notorious scavengers who would delight in picking our bones clean, these birds are not held in very high esteem by humans. But calling these cartoon predators "buzzards" is not only erroneous but a slight to a more amiable bird, one whose rightful name of "buzzard" has been pinned on a bird famous for this repugnant behavior, the vulture. Unfortunately, this misidentification is quite common: CNN (Cable News Network) once reported, "The condor is actually a type of buzzard."[1] It is not. And the citizens of Hinckley, Ohio, who admittedly know better, prefer to call their celebration of the annual return of the turkey vulture, "Buzzard Sunday."

Compared to the vulture, true buzzards are much smaller and more attractive birds. Better known in North America by "hawk" names, those common to the continent include the broad-wing hawk, red-tailed hawk, rough-legged hawk, and ferruginous hawk. Whereas this use of "hawk" is common in North America, in Europe such birds would be called by their proper name, "buzzard."

The name "buzzard" derives from the same Old English and Old French root as its genus name, *Buteo.* In ornithology, "buzzard" is commonly limited to the 26 raptor species of *Buteo,* although it is sometimes applied to all 90 species of the *Buteo* subfamily Buteoninae. Those of *Buteo* are high-soaring, keen-eyed birds of prey that wheel about in wide circles, looking for a victim. Spotting their quarry, they immediately drop out of the sky in a steep dive to pick off their targets—often small rodents. If ground food is scarce, a buzzard will readily attack other birds in the air. Yet this quick action belies its normally lazy demeanor. When not at hunt, it will spend hours idly riding warm air currents.

Most buzzards do not hunt the same territory year-round but migrate with the change of seasons. Of all the raptors, they are best known for their large migrating flocks, which typically follow major terrain features and weather patterns in their journey. Arriving at a brooding ground, they make their nests out of sticks and twigs, often in treetops or ledges 20 to 70 feet above the ground. Depending on the species, they will lay from one to seven eggs.

Buzzards range from 11 to 25 inches in length, with wingspans averaging 45 inches. Vultures, particularly those of the Old World, are much larger; their body lengths range from 25 to 45 inches, and their wingspans average 109 inches. (The largest of the vultures, the condor, has a wingspan close to 10 feet.) The typical buzzard has a heavy body, a short tail, and broad wings with somewhat blunt tips. Its underside is generally a mottled or streaked light color, whereas the mottling on its top side is most often a shade of dark brown. In contrast, vultures are not mottled but usually display bodies of a solid dark color (usually black or brown), with wings of either a single color or a pattern of light and dark areas. The buzzard has a typical hawklike head, with a heavy neck, a flattened crown, and a short, sharply curved bill. The neck of the vulture is much longer and in some species approaches half its body length. Likewise, its bill is quite extended, often comprising half its total head length. Because vultures seldom seize live prey, their talons are duller than those of buzzards and their feet weaker.

Unquestionably, vultures are not buzzards. They don't look like buzzards, nor do they act like them. So why do we persist in calling them buzzards? Perhaps it's the name itself, "buzzard." The double "z" sound, remindful of "gizzard" and "lizard," seems somehow fitting for a bird of scurrilous reputation. Actually, it is only North Americans who are guilty of this misuse; most everyone else in the world calls them by their proper name. The origin of this American "fowl-up" has been traced to the New World's early European settlers. Spotting the high-flying black vulture of the South making tight circles in the sky, they mistook it for the *Buteo* of Europe and called it a buzzard. Unfortunately, the term soon became a common misnomer that has persisted to this day.

The black vulture and the turkey vulture are North America's two principal vulture species. A third, the king vulture of Central and South America, will sometimes make a brief appearance in the southern border states. Altogether, the New World vulture family Cathartidae consists of seven species, including the largest bird of prey in the world, the condor. All are thought to be related to the stork and the flamingo.[2] The remaining vultures of the world, the Old World vultures, belong to the Accipitridae family, which also harbors all true buzzards. The 15 vulture species of Accipitridae arose separately from the Cathartidaes, yet through parallel evolution they developed remarkably similar features. New World and Old World vultures have similar heads, bodies, wings, and elongated beaks. They all delight in soaring for hours and share identical feeding habits.

Among their shared features, the most distinctive is a bare or nearly bare head attached to a typically long neck. This is not a trivial peculiarity but a very practical attribute. Unlike buzzards, vultures seldom pursue prey. They may hover above an intended meal or sit and watch it from above, but they will usually wait for its death before approaching. This is their preferred meal, carrion, often the result of a killing made by another animal, or sometimes a rotting carcass that may have been too tough to tear into immediately after death. With their extended necks and long hooked beaks, they will plunge their heads inside a carcass, searching for soft tissue, and in the process work their heads around in the viscera. Lacking head feathers, they avoid collecting the blood and gore that would accumulate on feathered heads. But even so, their feeding habits usually make them smelly and unpleasant birds to be around. Besides carrion, vultures also feed on small mammals, as well as on birds, fish, shellfish, and vegetation.

Despite the similarities of the two vultures, evolving independently of each other has produced significant differences. The feet of the New World vulture tend to be less strongly hooked than those of the Old World vulture, and the first joints of the inner toes are elongated. Like most birds, each Old World vulture communicates with a strong voice, but no New World species, strangely enough, has a syrinx (a voice box), and these birds can only make croaks and hisses. The New World vulture also has round eye orbits and open slit-like nostrils, whereas Old World species have eyes shaded by bony ridges and have round and partitioned nostrils. Compared to the Old World vultures such as the griffon of Africa, the North American vulture (except for the condor) has a relatively short neck and does not exhibit the crook or the bend in the neck as do the others—a feature often portrayed in vulture caricatures.

Vultures are master gliders, especially the turkey vulture with its 6-foot wingspan. Riding air currents, this bird migrates between southern Canada and Central America in flocks of 20 to 30 birds. Unlike the black vulture, it does not wheel about in the sky waiting for prey but prefers to linger on a high perch to watch for its dinner—an image sometimes portrayed in cartoons, most notably in Thomas Nast's famous depiction of Boss Tweed. The black vulture rarely migrates but instead keeps to Mexico and the southern and eastern seaboard states. All vultures scavenge in the country unless lured to towns and cities by poor sanitation practices, which are most common in underdeveloped countries. Like the buzzard, the Old World vulture will build a nest of sticks and twigs in trees or rock crevices to lay its one or two eggs. Curiously, the New World vulture does not bother with nest building but will lay its pair of eggs on the bare ground, on rock ledges, in caves, or in abandoned buildings.

Hawk • Falcon

Most of the confusion that has arisen over "hawk" and "falcon" in North America is attributable to the casual (and often unintentional) misuse of the two terms. Three practices in particular share much of the blame. The first is the inadvertent misnaming of daytime raptors by early American colonists. Unfamiliar with the differences in various birds of prey, these settlers mistakenly used the term "hawk" for a number of falcons. They called the peregrine falcon a duck hawk, the merlin a pigeon hawk, and the kestrel a sparrow hawk. To some extent, these misnomers are still with us.

The second questionable practice arose out of falconry. This "sport" is said to have originated in the Near East about 4,000 years ago. Eventually finding favor among Europe's upper class during the Middle Ages, it was a pastime that used not only falcons for hunting but other raptors as well. Persons of nobility commonly adopted the most spectacular and ablest raptor that their station in life would permit, with the king likely to reserve the gyrfalcon or, in some cases, the more majestic eagle for himself. The fifteenth-century *Boke of St Albans* lists the following dictate:

Emperor	Eagle, Vulture, or Merloun
King	Ger Falcon and the Tercil Ger
Prince	Falcon Gentle and Tercel Gentle
Duke	Falcon of the Loch
Earl	Falcon Peregrine
Baron	Bustard
Knight	Scacre and Sacret
Esquire	Lanere and Laneret
Lady	Marlyon
Young Man	Hobby
Yeoman	Goshawk
Poor Man	Jercel
Priest	Sparrowhawk
Holy Water Clerk	Musket
Knave or Servant	Kestral

Source: Glasier 1989, 19

Those birds toward the bottom of the list are the less dynamic birds of prey, which is why they were left for commoners. These raptors were not used so much for sport as they were a means of providing needed food for the table. To distinguish this utilitarian use of raptors from the sport of falconry, the former was often referred to as "hawking." With the advent of firearms, the sport eventually began to wane among the aristocracy and fell out of fashion. At the same time, the nobility began to loosen its grip on its subjects, allowing many hawkers to rise in status. Whether they at last felt closer to nobility or they just wanted to appear that way, hawkers began using the term "falconry" for their pastime, even though the birds they used were not falcons. To be sure, the higher the status these men attained, the more apt they were to choose one of the more prestigious hawk species—a practice mirroring that of the nobility.

Although all of these birds may now qualify for the sport of falconry, the falcon remains the preeminent sporting raptor. But some purists renounce this inclusion of nonfalcons within the sport falconry, preferring instead the term "hawking" to describe the use of any raptor other than a falcon.

The third dubious practice contributing to the confusion between the two is the broad use of "hawk" that is often found in books and articles. Books such as *Hawks in Flight* and *A Field Guide to Hawks,* part of the

Peterson Field Guide Series, cover not only hawks but eagles, falcons, kites, ospreys, and vultures as well. Including these other birds under the umbrella term "hawk" is usually a convenient way of indicating "diurnal birds of prey" (daytime fliers) and is not meant to imply that they are true hawks. Unfortunately, this misleading application occurs not only in the titles of books but in their text as well. But regardless of the term, the broad use of "hawk" and "diurnal birds of prey" is meant to exclude owls, those delightful nocturnal raptors.

Taxonomically, hawks and falcons belong to two separate families, Accipitridae and Falconidae, both part of the order Falconiforme. The hawks (specifically those of the genus *Accipiter*) share Accipitridae with eagles, buzzards, kites, ospreys, and Old World vultures. The falcons and the caracara make up Falconidae.

Unlike the British, who restrict the use of "hawk" to only those birds of *Accipiter,* North Americans use the term a bit more loosely. While normally excluding eagles, harriers, kites, and ospreys (though some would include the harriers), the North American use of "hawk" often includes not only the "true hawks" of *Accipiter* but also the buzzards of *Buteo* (closely related hawklike family members), plus various "buzzardlike hawks" of other genera living outside North America. (Buzzards can be easily distinguished from those of *Accipiter* by their chunkier bodies and longer and deeper wings.)

Hawks and falcons are about equal in size, both ranging between one and two feet in length. Their powerful feet have long sharp talons, and their beaks are sharply hooked and pointed. In flight, the most noticeable difference between the two is their wings. Hawks carry broad, deep wings that end in a rounded display of feathers splayed apart like fingers, whereas the wing of the falcon is more streamlined. The falcon wing, not as deep, is pointed at the end and its trailing edge sickle shaped. Unlike the buzzards, falcons and the smaller-winged hawks intersperse their gliding with rapid wing beats and do not soar unless migrating.

Other than variations in their wing feathers and molting patterns, the principal differences between the two lie in the head features. The falcon has a small tooth-like projection on each side of the cutting edge of its upper mandible that helps in breaking the neck of its prey; hawks have no such feature. The hawk also lacks a tubercle, a small projection found in the falcon's nasal cavity that is thought to aid in high-speed flight. Finally, bristles cover the cere, or nostril area, of the falcon, whereas the ceres of most hawks are bare.

Many of these physical variations translate into behavioral differences. With a sleeker profile, the falcon is much faster in pursuing prey than is the hawk. Because falcons reach estimated speeds in a dive (called stooping) of 100 to 175 miles per hour, theoretically, there is no animal or bird that it cannot catch. Unlike the impatient hawk, the falcon will take time to set itself up in the best possible position from which to strike. Its method of killing is quick; in usually less than a minute after making a capture, the falcon will have broken the cervical vertebrae of its victim. The hawk, in contrast, is less aggressive and efficient. Other than the mis-named hawks of *Buteo,* hawks rarely stoop from great heights; instead, they prefer to attack from a high perch or a low swoop. Unlike the for-tunate prey of the falcon, that of the hawk often suffers a prolonged death as it is plucked apart. And like the vultures in its family, hawks are not above dining on carrion, a meal that falcons pass up.

Falcons, with their ability to spot prey from great distances and then be upon it in a flash, favor the open country with its wide target area. Buzzards, because of their predilection to soaring, also fly the same types of open sites. Hawks hunt in more wooded locations, where perches are available and their deep wings allow them to more carefully maneuver among the trees. This is where they will also make their nests, usually in trees, on the ground, or on adjacent cliffs. In contrast, the falcon is not a nest builder. It may take over an abandoned nest and perhaps improve or repair it, but it may just as likely lay its eggs on a bare ledge or in a tree hollow but not on the ground. If one is unsure whether a captured or a tamed bird is a hawk or a falcon, get it excited (don't ask how); the hawk will wag its tail from side to side whereas the falcon will bounce its tail up and down.

Crane ● Heron ● Egret

Like many similes and metaphors adopted from bird lore—"eyes like a hawk" and "goose stepping"—the expression "to crane one's neck" is well chosen. One of the identifying characteristics of cranes is the way they keep their necks fully extended during flight. Swans, storks, and flamingos also fly with their necks stretched out, but none appear to be straining any harder than the crane. In contrast, the long-necked herons and egrets appear to be more relaxed when flying. Because of the unique shape of the sixth cervical vertebra, their necks fold back in tight S-con-figurations, putting their heads next to their bodies.

The heron is well known as a large bird with elongated features. Its long bill, neck, body, legs, and toes all contribute to the appearance of a bird having been subjected to a medieval stretching rack—the same might also be said of the crane. Yet some heron species scarcely come close to this description. The green-backed heron and the black-crowned night-heron, which are the size of crows and ravens, have necks no longer than do these two black birds. They do somewhat resemble the larger herons with their elongated shapes and spearlike beaks, but their smaller size and short legs make them easy to misidentify. However, they each have two to three unique paired patches of powder (down) feathers, as well as comb-like inner margins of their middle toes, which readily identifies them as herons.

As wading birds, herons share their family Ardeidae with the egrets and the bitterns, birds usually considered to be types of herons. Like the green-backed heron, bitterns are smaller, medium-sized birds. The egret, which closely resembles the long-legged herons, has striking snow-white plumage (which also occurs in some immature heron-named species), especially when displaying its unique breeding feathers, called an aigrette (the source of its name). More a curse than an enhancement, this delicate and loosely attached plume that streams down the egret's back was once responsible for the near extinction of several species. During the nineteenth century and into the early part of the twentieth century, women's fashion created an incredible demand for these lacy plumes. To meet the demand, hunters so relentlessly pursued egrets that their killings quickly became acts of wholesale slaughter. Fortunately, public outcry, hastily enacted government regulations, and the inevitable change in fashion saved these birds from a dire fate.

Herons are closely related to three other large wading birds within their order, storks, ibises, and flamingos. Like many of these, the heron carries long, rounded wings; each foot has four long toes that are useful in supporting the bird on soft mud; grows a long, tapering bill; and sports a short, square or rounded tail.

The cranes (family Gruidae), occasionally confused with herons, are very similar in several ways but are more closely related to the rails and the bustards, much smaller birds. The crane has long wings similar to those of the heron, but unlike the heron, its hind toes are small and elevated, leaving it with only three toes per foot for support. Its bill is also shorter, and its tail ends in an assemblage of feathers that droops over its rear like a bustle. In contrast to the looser feathers of the heron, which impart a somewhat ruffled look, the plumage of the crane is much denser and more compact, giving its body a smooth appearance.

Two other, less apparent, differences involve the toes and the windpipe. Unlike those of the crane, the middle front toes of the heron end in a claw called a feather comb. Toothed like a comb on its outer edge, it is used for scratching, preening, and removing parasites. The windpipe of the heron, despite the kinked shape of the neck, allows air to move in and out unhindered, whereas the crane has such a convoluted windpipe tucked behind its breastbone that when it forcefully exhales, it emits resonant notes audible for several miles. The whooping crane is well named after this ability to let go with a loud, alarmlike *kerloo!*

As with their physical features, the behavior and the activities of herons differ considerably from those of cranes. In flight, herons flap their wings in strong, regular beats, whereas cranes will make a slow deliberate downbeat, followed by a quick upstroke. Coming in to land, the heron will occasionally set down in a tree, a favored nesting spot. But some species find other nesting locations equally as inviting; the great blue heron is just as likely to build its nest on the ground or on the rocky ledges of cliffs. Cranes are strictly terrestrial birds that never sit in trees, preferring instead to make mounded nests of grass in isolated wet areas, which can be anything from marshes to wet grasslands. Grasslands are also attractive feeding spots for the crane. Savannas, pastures, and even dry uplands complement the more favored shallow-water feeding grounds. Omnivorous in their diet, cranes feed on grains, leaves, fruit, and roots, as well as insects, frogs, small rodents, and various aquatic life, but rarely will they go after fish. Herons stick closer to wetlands, favoring lakes, ponds, rivers, marshes, and seashores. Accordingly, their diet consists more of aquatic foods such as crustaceans, frogs, lizards, fishes, sedges, and water lilies. Yet herons are not above feeding on grass seeds, small rodents, and even other small birds.

Grouse ● Ptarmigan ● Quail ● Partridge

In Robert Elman's *The Hunter's Field Guide to the Game Birds and Animals of North America*, six alternative names are listed for the white-tailed ptarmigan. Included in this list are: snow *grouse*, snow *partridge*, and mountain *quail*. If this mix of names appears a little peculiar, grouse are known by an even greater assortment of names. Such variations in the names of game birds and animals are not at all uncommon: hunters seem prone to casually name their quarry as they see fit, either making up new names or misapplying old ones—consider the poor mountain lion, with its list of aliases (see "Jaguar ● Cougar ● Leopard ● Panther"). For the white-tailed ptarmigan, the alternative name "snow grouse" is certainly erroneous, but it is not as far off the mark as are the labels "partridge" and "quail."

All four birds in question belong to the pheasant and turkey order Galliformes: the partridges and the quail making up the family Phasianidae, and the grouse and the ptarmigans constituting the family Tetraonidae. The chickenlike Tetraonidae species have short, rounded wings and short, curved beaks and spend most of their time on the ground, trying to scratch up a meal—most fittingly, the prairie-chicken, a grouse, was named for this poultrylike behavior. Grouse and their kin differ from the others in their order by shrouding their nostrils with feathers and sporting feathers on their legs that, in most species, extend down to the base of their toes. In addition, in the winter grouse develop comblike teeth on the sides of their toes, which they use for grooming. Some males also bear inflatable "air sacs" on their chests to attract females. Except for the ptarmigans, members of the grouse family tend to be relatively weak fliers. This deters them from migrating too far, if at all. Keeping to the more northern temperate climates, they inhabit a variety of regions, including tundra, forests, grasslands, and desert scrub, where they survive by digging up insects and herbaceous ground food.

Although ptarmigans are grouselike, they do make up their own genus, *Lagopus* (grouse occupy several genera). The name *Lagopus* (meaning "rabbit foot" in Greek) describes one of their more prominent features, their feather-covered feet. Ptarmigans also stand out because they do not grow "combs" on their toes in the winter. These strictly arctic birds are smaller than the grouse (average weight of a male: 18 oz. versus 40 oz. for a grouse) and comparable in size to the partridge. Unlike any of the three other birds, the ptarmigan goes through one extra molt each year and has plumage dramatically different in winter than in summer. Because the ptarmigan lives on snow-covered ground in the winter, its white plumage allows it to blend into the background; likewise, its patchy brown appearance during the summer acts as camouflage.

Why the white-tailed ptarmigan has been called a partridge and a quail is anyone's guess. Partridges and quail may resemble each other, but neither looks like a ptarmigan. As with the ptarmigans and grouse, quail have short wings and short sharp beaks, yet overall they are much smaller birds, usually under a foot in length (grouse are about half again as large). Neither the quail's legs nor its nostrils have feathers, and its head, sitting on a shorter neck, most often sports a feathered plume from its crown. The quail of Europe, unlike their North American counterparts, have no head plumes, and they never grow more than 6 or 7 inches in length. Also peculiar to the quail is a slight indentation on the cutting edge of its lower mandible. Similar to grouse, quail scratch the ground for food, like chickens do, but their diet consists more of grains than herbs. Except for the mountain quail, North American quail prefer warmer regions, where they range from the grasslands of the north to the cloud forests of Mexico.

Four species of partridges live in North America—no more, no less, despite the many ground species (the ruffed grouse, for one) that have been erroneously tagged with the name "partridge." All were introduced from the Old World. The gray partridge and the Chukar partridge are the two dominant species, with the Barbary and the red-legged partridges scattered in a few western states. Superficially, the partridge is very similar to the quail but differs taxonomically in its lack of an indentation on its lower mandible. A little larger than the quail, the partridge also may be distinguished by its lack of the characteristic quail head plume (some quail, such as the northern bobwhite and the Montezuma, also lack a plume). One unique feature borne by some partridges is a horny covered outgrowth of bone on the rear inner side of the legs, called a spur, a growth that is absent on quail.

Gull • Tern

Many people, especially those living inland, regard most medium-size seabirds as gulls. The close relationship that gulls have with the sea has even fostered the notion that all gulls are *sea* gulls and that "seagull" is a proper, all-inclusive name, even for those species living in mid-America, such as the Franklin's gull. Coastal residents are generally a bit more gull savvy; while occasionally mistaking a particular bird for a gull, they are usually aware that petrels, fulmars, shearwaters, and prions are only gull-like and not true gulls. But whether spotted by a landlubber or a sailor, no bird is more commonly mistaken for the gull than its nearest relative, the tern.

The 43 species of gulls and 39 species of terns, considered to have evolved from a common ancestor, make up the family Laridae. Unlike the petrels, the fulmars, and the shearwaters, those of Laridae lack a breathing tube on the ridge of their bills, and unlike the prion, have no food-straining device on their upper mandible. Gulls and terns, some of which are called kittiwakes and noddies, are medium-sized, long-winged birds with plumage ranging in color from white to black; they occasionally sport wings dappled in brown. Commonly seen at sea, few venture very far from land. For many of them, bays, marshes, and lakes are as welcome feeding grounds as the open sea is. But despite their close relationship and shared similarities, gulls and terns remain so conspicuously different that any confusion arising between the two should be minimal.

Gulls are robust birds, often appearing somewhat gray or dirty looking from a distance. They stay aloft by gliding and soaring, with occasional assistance from slow-flapping wings. The smaller, swallow-shaped terns do little gliding and fly with more of a snap in the beat of their wings. Unlike the tern's short inner wing limb, that of the gull is at least as long as the outer limb; and whereas the tail of the tern is definitely forked, that of the gull is usually rounded. Up close, the tern can be readily distinguished from the gull by two features of the head. Most terns have black "caps," which on some species form crests. They also carry fairly long, narrow, and sharply pointed bills, which in flight are generally pointed downward. Gulls have no such head markings, and their bills, normally held straight ahead when flying, are heavy and appear slightly hooked at the end in large species or are short and pointed in the smaller species.

Gulls are omnivores that delight in feeding not only on seafood but on garbage, carrion, and even the eggs and the young of other birds. Add to this food list sewage, rodents, hares, insects, earthworms, and grain, and it is not surprising that they survive as well inland as they do at sea. Over water, the gull will hover and then settle on the surface to feed from a floating position. To feed its young, the gull swallows an intended meal, returns to the nest, and then regurgitates it. The more slender and graceful terns prefer to swoop down and snatch food from the surface or plunge underwater if necessary. Much more selective in their dining, they will eat only live food, which can be anything from fish and squid to shrimp, plus such land food delicacies as insects and worms. Unlike gulls, most terns carry food back to their young in their bills.

Gulls are very gregarious birds, sometimes assembling into flocks numbering in the thousands. When not in flight, they spend a good deal of their time on water, paddling around or snoozing between meals. The more slightly built terns rarely swim at all, preferring to rest along the shore or on floating debris. Neither do they gather in large flocks, except perhaps at nesting time.

Pigeon ● Dove

Of the approximately 295 pigeon species in the world, only 16 reside in North America. The most common New World species is the rock pigeon *(Columbia livia),* which can often be found taking up squatter's rights in our towns and cities. This plumpish, usually gray, 13- to 14-inch-long bird is one of the few pigeons that is comfortable around man, an affiliation possibly stemming from the breeding their ancestors underwent years ago by pigeon fanciers—the rock pigeon was first introduced to North America in 1606. Often considered the "typical pigeon," it is sometimes called a rock dove, an equally valid name. In fact, three-quarters of all North American pigeon species are better known as doves, and those that are not could rightfully claim the name.

"Dove" is no more valid a term than is "pigeon"; it is only an alternative. But because "dove" is associated with a handful of similar-appearing species, setting them apart from those known as "pigeons," it is often assumed that the two terms truly signify taxonomically different birds. And although the species we commonly call pigeons are on average larger than those we call doves, size is not a credible criterion for distinguishing these two artificial groups.

Neither is tail shape, a feature often used to distinguish the common pigeon from the mourning dove, which is North America's second-most common member of the pigeon family. The tail of the feral rock pigeon is broad and rounded, whereas that of the mourning dove is long and tapering—a feature that varies greatly throughout the family. It also has been observed that pigeons tend to be more arboreal than doves, but again, this characteristic is not important enough to separate the two.

As may be guessed, other than the use of "dove" and "pigeon" in names long established by tradition, science lends little weight to the two terms and considers them interchangeable. Although ornithologists regard "pigeon" and "dove" as synonyms, in the everyday world these terms do carry different connotations, the origins of which go back hundreds of years.

Up until the eleventh century the English called all members of the Columbidae family "doves," or *dufes.* The word derives from the Dutch *duyve,* which is akin to the Danish *due* and the Icelandic *dufa*—all meaning "a diver." At the same time, the French and the Italians were calling these birds *pijons* and *picciones,* both terms coming from the Latin *pipire,* meaning "to peep." With the Norman invasion of England in 1066, the term *pijon* was introduced to the British Isles in association with Scandinavian/French cooking. For the English, the economy and ease of using the larger doves for their adopted French dishes eventually led to the use of "pigeon" for such species, while "dove" continued to be used for all others. Today, the trend in English cuisine has been to replace "pigeon"—a name having since acquired somewhat unappetizing connotations—with the more appealing "dove." And when the bird is served up, more often than not it will be called squab (a Scandinavian word meaning fat flesh), a term also used to denote a very young pigeon.

Borrowing a foreign term for an animal when used as food and retaining the native term when not is not at all unusual. Two other French terms—"veal" for calf, and "escargot" for snails—are used when these foods are served for dinner. Likewise, the Latin word "calamari" is used for squid; and "caviar," a term originally from Turkey, for roe (fish eggs). The continued use of pigeons as food eventually imparted a utilitarian connotation to the word, which ultimately found its way into everyday expressions. We use "pigeon" for a fool or a dupe, an informer (stool pigeon), feet turned inward (pigeon-toed), and a compartment or an act of deferment (pigeonhole). Even the Bible speaks of sacrificing pigeons, not doves; yet the bird Noah sent out to find land was called a dove—a harbinger of salvation. And this is typical of the more virtuous connotations "dove" has acquired. We use it as a symbol of peace, an affection for a loved one, an icon of the Holy Ghost, and in general an expression of purity, gentleness, and innocence. It is remarkable that these two terms, one used for the mundane and the other for quite lofty notions, refer to the same bird.

7

MAMMALS

Mammal ● Marsupial ● Monotreme

Through carefully dated fossils, we know that 215 million years ago the dinosaurs were well on their evolutionary march toward becoming the most awesome animals on earth, a reign that would last 150 million years. However, these beasts were not alone. At the very same time the dinosaurs were populating the landscape, mammals were also evolving, but because dinosaurs dominated the world, mammals were relegated to playing only a minor part in Earth's developing zoo. It was not until the dinosaurs went into their disappearing act—about 67 million years ago—that mammals, first emerging as small, shrew-sized creatures, were able to evolve into more prominent forms. No longer kept in check by the big sauropods and their kin, mammal evolution produced a wealth of animals that now numbers close to 4,250 species.

Like the reptile forms from which they evolved, all early mammals were egg layers, which zoologists call prototherians (meaning "first beasts"—*therian* coming from Greek for "beast" or "animal"). Never producing a great variety of species, their direct descendants, the monotremes (classified as ornithodelphians—meaning "born like birds"), barely evaded extinction as they made their way into the present era. But despite their role as a minor animal group, this ancient line of egg layers did successfully represent mammals for over 120 million years. Then, about 100 million years ago, a new type of mammal began to emerge, one that bore live young.

During the ensuing 25 million years, these live-bearing mammals slowly evolved into two major forms: the metatherians ("subsequent beasts"), most of which developed external pouches for their

underdeveloped newborn; and the eutherians ("good beasts"), which
gave birth to well-formed young. Although we commonly call the eutherians "mammals," the metatherians "marsupials," and the egg-laying
ornithodelphians "monotremes," all three are true mammals.

The label "mammal" simply indicates these animals all share and are
united by a very important feature not found in any other group of animals: mammary glands. These glands supply the newborn with its first
source of nutrition, milk. Whereas almost all mammals use nipples to dis-
pense their milk, the monotremes lack such a delivery device and simply
secrete milk from modified sweat glands onto their belly fur, from where
the young lap it up.

Nature, then, has left us with three distinct forms of mammals. The
monotremes, represented by just 6 species in 2 families: the duckbilled
platypuses, and the echidnas or spiny anteaters; the marsupials, with
about 250 species divided among 8 families; and the eutherians, better
known as placentals, with close to 4,000 species grouped into 22 families.

Monotremes get their name from the design of their urinogenital systems. In almost every placental, such as ourselves, the anus, the urethra,
and the vagina have separate openings. In a monotreme, however, all
three passages merge to form a single ventral opening called a cloaca,
hence the term "monotreme" (*mono:* "one," *treme:* "aperture"). But this
feature is not absolutely unique to the monotremes. The pika and the ten-
recs, placental mammals, also have such singular openings.

After delivering its leathery-shelled egg (similar to that of a reptile), a
monotreme performs incubation in one of two ways. Before laying her
eggs, the duckbilled platypus excavates a burrow in which she builds a
nest. After laying one to three eggs, she will keep them warm beneath her
body until hatching, typically a 2-week task. The echidna, however,
makes no nest. It incubates its lone tiny egg next to its abdomen in a flap-
like pouch made of belly skin. The newly hatched echidna will remain in
the pouch, lapping up milk until old enough to survive on its own.

The marsupial is so called because of its marsupium, an external
abdominal pouch it uses to shelter its newborn (in some species, this
pouch forms only during pregnancy and in other marsupials not at all).
The newborn—really a blind and naked underdeveloped fetus—is not
born in the marsupium but must squirm and claw its way from the birth
canal to this pocket nursery where, finding a teat on which to fasten, it
will remain until it matures. Before this occurs, however, in its first stages
of development the marsupial begins as an egg, not like the typical eutherian egg that attaches to the uterine wall but one with a rich yolk sac similar to that of a reptile or a bird. As a source of nourishment equivalent to

the eutherian placenta, the egg sustains the developing marsupial for its 10- to 30-day stay in the uterus. Although this is true of almost all marsupials, the bandicoot embryo develops in the uterus while attached to a placenta. Even though the bandicoot placenta is relatively well formed, it is not particularly large and, oddly enough, remains connected to the embryo by the umbilical cord as the newborn climbs into the marsupium and attaches itself to its mother's teat. A koala embryo develops in somewhat the same manner; however, the connecting placenta is neither as fully developed nor as functional.

Placental mammals are named for the vascular organ within the uterus that supplies the developing fetus with oxygen and nourishment and that also accepts fetal waste. Unlike the developing marsupial, the placental remains in the uterus considerably longer (19 days from conception for a mouse and 624 days for the Indian elephant, versus 8 days for the opossum, a marsupial). Maturing within the protection of its mother's womb, a placental mammal has no need for a pouch after birth.

While mammary glands tend to be their most noted feature, these three mammal forms share other major identifying characteristics. Most obvious is their body hair, although on some, such as whales, it may be no more than a bristle or two above the upper lip. Another feature, not at all as apparent, is the single bone that forms the mammalian lower jaw—except for the jawless fish, all other vertebrates have multibone jaws. Other unique features include the complex crown pattern of the cheek teeth, three bones (hammer, anvil, stirrup) within the middle ear, and sweat glands. Some features, such as a four-chambered heart and warm-bloodedness, commonly thought to be unique to mammals, actually occur in other animals. Birds have both characteristics, and a few fish, such as the porbeagle and the mako (both sharks) and the tuna, also regulate their body temperatures. In fact, certain tuna species maintain body temperatures of at least 18° F above that of the water. It has also been well argued that dinosaurs were warm-blooded creatures. Conversely, the naked mole rat has recently been found to be a cold-blooded mammal.

As the most primitive of the mammals, the six monotreme species, besides possessing single ventral openings and laying eggs, have retained several other reptilian characteristics. For example, the skeletal design of their pectoral shoulder girdles—in part, the use of a T-shaped bone to connect the collarbone to the sternum—is more reptilian than mammalian, leaving them with a somewhat reptilian posture. A monotreme also has cervical ribs, an only slightly coiled cochlea (the spiral structure of the ear that contains the auditory receptors), an imperfect atrioventricular valve in the heart, and electroreceptors on its snout—all unique features. A monotreme also lacks lachrymal bones (small bones forming the inner wall of the eye socket), an almost universal feature among the other mammals.

Once a far-ranging animal, monotremes now occupy only a small region of Earth. Confined to Australia, Tasmania, and New Guinea, the platypus and the echidna have evaded extinction only because of an almost total lack of competition from other animals. The platypus, a semiaquatic furry carnivore with webbed feet, a ducklike bill, and no teeth (except when young), feeds on freshwater fish and invertebrates. The echidna is quite different. It is more a terrestrial animal, with a body covered in sharp spines; its face tapers to a long snout from which a long sticky tongue flicks out to nab insects and worms. To secure a meal, the echidna uses powerful claws to dig for ants and termites. Unique to both monotreme males is a spur on each hind ankle. It serves no apparent purpose on the echidna, but on the platypus it connects to a venomous gland whose poison can be extremely painful to humans and deadly to small animals.

Many marsupials, while not as peculiar as the monotremes, are nonetheless odd-looking creatures, particularly the kangaroos, the quokka, and the wombat. Yet quite a few of the 248 living species closely resemble placental mammals—both forms having adapted to similar environments in similar ways. The following marsupials and their placental lookalike counterparts reflect the variety that such convergent evolution has wrought.

Marsupials	Placentals
Marsupial mole	Common mole
Cinnamon antechinus	Kangaroo rat
Gunn's bandicoot	Bachman's shrew
Fluffy glider	Southern flying squirrel
Mongan ringtail	Sportive lemur
Koala bear	Bear cub
Tasmanian wolf	Several midsize shorthaired dogs

Outward appearances aside, all marsupials share specific features that distinguish them from placentals.

- A marsupial has a double lateral vagina and usually a two-lobed penis; a placental has a single vagina (although it may be divided) and a single-lobed penis.

- The marsupial braincase and its cerebral hemisphere are usually smaller than those of the placental.

- Marsupials have more incisors and molars but, typically, fewer premolars.

- Unlike the pelvic girdle of placentals (the bony arch supporting the hind limbs), that of marsupials (and monotremes) contains epipubic bones. These two large, flat bones flare forward and outward to support the abdominal muscles and the marsupium.

Although Australia is well known for its marsupials (except for bats and rodents, marsupials and monotremes are the continent's only indigenous land mammals), the New World also has its share. At least 72 species of marsupials, consisting of opossums, shrew-opossums, and the colocolo, live between Chile and Ontario, Canada. Other marsupials inhabit Tasmania, New Guinea, and the Celebes—none live in Asia, Africa, or Europe. The placentals, residing in all but a few climatically extreme niches on Earth, make up the remainder of the mammal spectrum. From the smallest, the Etruscan shrew weighing in at 0.09 oz., to the largest, the 150-ton blue whale, they have adapted to just about every form of environment the planet has to offer.

Evolutionary forces have encouraged placental mammals to diversify into an amazing variety of creatures. Some, like shrews, moles, anteaters, and sloths, remain quite primitive, whereas the primates stand near the forefront of evolution (taking intelligence and dexterity as the ultimate measures). Between these extremes, the placentals make up a true menagerie of opposites: meat-eating hyenas versus plant-nibbling rabbits versus insect-devouring shrews versus omnivorous humans. Winged bats versus flippered seals versus long-fingered tarsiers. Arboreal lorises versus burrowing prairie dogs. Diurnal gorillas versus nocturnal hedgehogs. Swift pronghorns versus sluggish sloths. Ferocious wildcats versus docile guinea pigs. Virtually hairless whales versus furry polar bears versus scaled pangolins, and so on. It's a remarkable listing but nowhere complete. In fact, the diversity of features among placentals seems almost excessive. Simply consider the following list of placental mammals still roaming Earth.

Order	Example
Chiroptera	Bats
Primates	Lemurs, tree shrews, lorises, marmosets, monkeys, apes, humans
Insectivora	Shrews, tenrecs, moles, hedgehogs
Pinnipedia	Seals, seal lions, walrus

continued on next page

continued from previous page

Order	Example
Edentata	Anteaters, armadillos, sloths
Pholidota	Pangolins
Lagomorpha	Rabbits, hares, pikas
Rodentia	Mice, rats, squirrels, beavers, chipmunks, porcupines
Cetacea	Toothed whales, baleen whales, porpoises, dolphins
Carnivora	Dogs, cats, bears, raccoons, weasels, mongooses, hyenas, seals, walrus, sea lions
Tubulidentata	Aardvark
Proboscidea	Elephants
Hyracoidea	Hyrax
Sirenia	Dugong, manatees
Perissodactyla	Odd-toed ungulates: horses, zebras, tapirs, rhinos
Artiodactyla	Even-toed ungulates: pigs, peccaries, hippopotamuses, camels, chevrotains, deer, giraffes, pronghorn, cattle, antelope, goats, sheep

African Elephant ● Indian Elephant

It is a common misconception that the mammoth was a predecessor of today's elephant. Fossil discoveries indicate that the two animals evolved independently of each other 5 million years ago during the Pliocene epoch. Although not an ancestor, the mammoth was the elephant's closest living relative for millions of years and relinquished that status to the dugong and the manatee only when it became extinct. Oddly enough, next to these sea creatures, the elephant's next closest kin may well be a small, sharp-toothed, guinea pig–like animal called a hyrax (though some zoologists have recently questioned the hyrax's close kinship to elephants). The most important features uniting these three disparate animals are shared protein factors and wrist bone configuration (their carpal bones sit directly above each other, whereas in all other mammals these bones are staggered). Together, the elephants (order Proboscidea), the dugong and the manatees (order Sirenia), and the hyraxes (order

Hyracoidea) make up the superorder Paenungulata (meaning "near ungulates"). In the vocabulary of science, hoofed animals are called ungulates, a word derived from the Latin *unguis,* for "nail." The nails of these animals wrap around their toes to protect and support their feet.

The name of the elephant order, Proboscidea, is well chosen: it refers to the elephant's trunk (proboscis) and literally means "before the mouth." "Elephant" is likewise a fitting term, made up of the Greek word *ele,* for "arch," and the Latin word *phant,* for "huge," a reference to its arched back. The outdated term for elephants, "pachyderm," heard years ago in the grandiose ballyhoo of circus barkers, comes from the old name of their order, Pachydermata, which referred to the elephant's thick skin.

Elephants come in two very distinct forms: one rightly called the African elephant, and the other suffering under the popular misnomer of "Indian elephant," which, although a valid term, really pertains to only one of three similar Asian subspecies. So a more practical way to look at the differences between the African genus, *Loxodonta,* and its counterpart is to focus not just on the Indian elephant but on the Indian elephant's genus, *Elephas,* a group better known as the Asian elephants.

Up until 2001 the African elephant was considered to be a single species, *L. africana,* which was distinguished by two subspecies: the bush elephant (also known as the plains or the savanna elephant) *L. a. africana,* and the forest elephant *L. a. cyclotis.* Now, because of new genetic evidence and other supporting factors, these subspecies appear to be two distinct species: *Loxodonta africana* and *Loxodonta cyclotis.*

The bush elephant, weighing between 8,800 and 15,400 pounds and standing 10 to 13 feet at the shoulder (the largest elephant in the world), is considerably larger than the 4,550- to 10,000-pound forest elephant that stands 6½ to 10 feet tall. The darker-skinned bush elephant also has proportionally larger and more triangular ears. Even more unusual, the forest elephant has 5 toenails on its front feet and 4 or 5 toenails on its back feet (a strange variation in itself), while the bush elephant's front feet may have 4 or 5 toenails and its back feet, 3, 4, or 5 toenails.

The elephants of Asia comprise a single species, *Elephas maximus,* which is divided into three subspecies: the Sri Lankan elephant, *E. m. maximus;* the mainland elephant of India, Indonesia, and Southeast Asia (the Indian elephant), *E. m. indicus;*[1] and the Sumatran elephant of Sumatra, *E. m. sumatranus.* With one exception, the differences among these three subspecies are not as significant as those separating the two African elephants. The Sumatran elephant, standing somewhat apart from the Sri Lankan and the Indian elephants, is decidedly unique in possessing 20 pairs of ribs, 1 more pair than the other two (African elephants have 21 pairs of ribs).

Aside from this anomaly, Asian elephants are quite alike. The Sri Lankan elephant is largest, ranging from 4,400 to 12,100 pounds, whereas the Sumatran, the smallest, weighs between 4,400 and 8,800 pounds. The Sumatran elephant, standing 10$\frac{1}{2}$ feet high at the shoulder, is about a foot shorter than the other two. It also has the lightest skin color of the three, although the Sri Lankan elephant, with the darkest skin, has large, unpigmented blotches on its belly, face, trunk, and ears.

An unusual order of mammals, African and Asian elephants are most noted for their trunks, flexible appendages accommodating the nostrils, which are constructed out of the union of their upper lips and noses. The trunk contains neither bone nor cartilage but is primarily made up of muscle and connective tissue. Other characteristics unique to all elephants include a large hulking body, long tusks, very thick skin, large floppy ears, and pillarlike legs. Not as obvious are their distinctive internal features, such as a heart with a double-pointed apex and tusks that bear a diamond-shaped growth pattern.

Also unusual is the manner in which elephant teeth replace themselves. Unlike most mammals, who replace an old tooth with a new one that pushes up from below, elephants have a continuous convoy of teeth pushing forward from the back of the jaw that displaces their older, worn front teeth. But just as these shared characteristics wed African and Asian elephants into a single family, their differences, nominal as they may be, help to divide the two into separate genera. Among these telling features, the most obvious relate to the head.

Without doubt, it is their ears that most easily distinguish the two forms. Unlike the smaller ears of the Asian elephant, the large ears of the African elephant extend above the top of the neck. Their trunks, while initially appearing very similar, can be distinguished by several variations. Most noticeably, the tip of the African elephant's trunk has two fingerlike appendages, one above the nostrils and one below, whereas the trunk of the Asian elephant has only an upper appendage. Because of their ability to pick up small coin-sized objects, these fleshy, maneuverable extensions are commonly called "fingers." The trunk of the African elephant also has more rings along its length and is somewhat less rigid than that of the Asian elephant.

Elephant teeth, unusual not only in their replacement sequence, also have a peculiar construction. Each molar and premolar consists of individual vertical plates (the maximum number in the African elephant is 14 and in the Asian elephant, 27), cemented together to form a single tooth. The enamel-grinding surface of these plates takes the form of loops. In the tooth of the African elephant, these loops have a lozenge shape (the source of the genus name *Loxodonta*) and generally touch one another. In the Asian elephant, the loops look more like a relaxed rubber band and never touch. The tusks of the African and the Asian elephants (their overgrown upper incisors) are very much alike, and it is only in their relative size that they differ. In the Asian elephant, the tusks of the female are greatly reduced, if not absent. In African elephants, both sexes carry well-developed tusks, although it is not uncommon for the female to have a somewhat smaller pair.

The shape of the cranium is also a distinguishing feature of the two. In the African elephant, the forehead and the top of the head are fairly flat, whereas in the Asian elephant, the forehead is dished and the top of the head has twin domes above the eyes. Because of its domed cranium and its habit of holding its neck at a 45-degree angle, the highest point on the Asian elephant is its head, whereas the highest point on the African elephant, which extends its neck horizontally, is its shoulders. To the rear of its shoulders, the back of the African elephant dips somewhat before rising at its haunches. In contrast, the back of the Asian elephant is convex or at least level along its length. At its rear, the African elephant carries a 3- to 4-foot tail that is supported by as many as 26 vertebrae. The tail of the Asian elephant, with 33 vertebrae, is decidedly longer, generally growing to between 5 and 7 feet in length.

Not quite as obvious a feature is their bellies. That of the African elephant slopes downward from the front legs to the rear, whereas the Asian elephant's belly usually sags a bit in the middle.

Vicuna • Guanaco • Alpaca • Llama

Although a few scientists still harbor doubts as to the place of origin of camels and their kin (animals collectively called camelids), most believe that the fossil evidence points to North America. Camels started out 40 million years ago as small rabbit- to sheep-size animals; it was through the continual expansion of grasslands that these emerging creatures were able to evolve into larger forms, most developing long legs and necks. By the Pliocene era (2 to 3 million years ago), many had begun moving into South America, where they gave rise to the guanaco (pronounced gwa-'nak-o) and the vicuna. The camelids choosing not to go south took two paths. Those finding their way into Asia over the Bering Sea land bridge between Alaska and Siberia became the forerunners of today's camels. Those remaining in North America did not fare as well and are thought to have perished during the ice ages. The two camel species, Arabian and Bactrian, which are often confused with one another, are discussed in their own section.

Until quite recently, all South American camelids (commonly designated lamoids) had been classified in one genus, *Lama:* the guanaco represented by the species *L. huanacos,* the llama by *L. gama,* the alpaca by *L. pacos,* and the vicuna by *L. vicugna.* Today, because of a better understanding of camelid development and a recognition of the vicuna's unique features, the vicuna has been assigned its own genus, *Vicugna.* Except for a change in the spelling of the guanaco's species label to *guanicoe,* the designation of the others remains the same. But despite their status as distinct species, the fact that lamoids can interbreed and produce fertile offspring calls into question their status as true species and further obscures their exact relationships. (Whereas the alpaca and the llama will voluntarily mate with each other, the alpaca and the vicuna will not—a trait commonly used to distinguish one species from another.)

Few authorities dispute the taxonomy of the guanaco and the vicuna: both animals appear to be reasonably distinct lamoids that began their separate evolutionary paths quite some time ago. For the llama and the alpaca, however, species that now exist only as domesticated animals, a growing controversy over their origins has raised questions as to their proper classification. Once they were believed to be the descendants of the guanaco, but certain physical and behavioral differences now seem to point to a different ancestor. Some authorities feel that features such as the alpaca's fine wool suggest a crossbreeding between the guanaco and the vicuna. But standing in the way of such likelihood are the vicuna's many other distinctive features, not the least of which is its size.

The vicuna, smallest of the lamoids, stands only 30 to 36 inches high at the shoulder, its head-to-tail length is about 5 feet, and its weight averages 110 pounds. In contrast, the guanaco and the llama, the largest of the lamoids, stand 4 feet high, are about 6.5 feet long, and weigh an average of 185 pounds (guanaco) and 300 pounds (llama). The alpaca, a considerably smaller species, is about 3 inches taller and 25 pounds heavier than the vicuna.

The vicuna, bearing a slender body, a relatively long neck, and a short head, is a delicate and graceful-appearing animal. And unlike its brethren, it does not develop "chestnuts": callosities on the inner side of its front legs. But more significantly, the vicuna's unusual teeth truly set it apart from the camelids of the *Lama* genus. Unlike all other artiodactyls (even-toed ungulates), the vicuna's lower incisors never stop growing, and like rodent incisors, enamel forms on only the face of these teeth.

Except for an unusual patch of long, silky, white hair on its throat and chest, from a distance the vicuna's coat of long tawny-brown or cinnamon wool appears rather unremarkable. But a closer examination will reveal that the vicuna has an exceptionally soft undercoat that is protected by very fine guard hairs. This coat is so luxurious that many commercial traders regard its wool as the finest in the world (it has brought up to $1,000 a pound on the open market). Endowed with wool so highly prized that it was once reserved for royalty, and cursed with a tasty flesh, the vicuna was hunted nearly to extinction. Numbering less than 10,000 in 1967, by 1990 subsequent conservation efforts raised their population to 130,000.

The vicuna, living in small groups consisting of either a male with several females and their young, or bachelor males, makes its home in the Andean highlands (primarily, Peru and the Altiplano of Bolivia), at altitudes generally between 12,000 and 18,000 feet. The grass and soft broadleaf herbs it grazes on usually lie within limited and fairly permanent feeding territories. This intentional limitation is necessitated by the vicuna's inability to store water, preventing it, unlike camelids of the *Lama* genus, from journeying more than a day's travel from water.

The guanaco, quite capable of storing water (even saltwater), is able to venture not only throughout the green meadows of the lowlands but also among the arid and semidesert highlands lying 14,000 feet above sea level. Such treks are also made possible by the guanaco's less discriminating diet, one consisting not only of grasses and herbs but also of shrubs, buds, and tree bark. Guanacos, by far the most widespread of the lamoids, range from northern Peru to Tierra del Fuego, with the vast majority living in Argentina. They travel in herds that vary from a male and four females to those consisting of fifty young bachelors. An aging guanaco's final trip usually ends at the herd's "dying ground." During the recent past, however, many never reached this final resting place, instead falling prey to hunters.

The unwelcome guanaco, once dominating the lush green meadows of Peru, had become the target of encroaching sheep ranchers. As an undesirable resident whose meat and pelt held only fair value, the guanaco suffered an unmerciful slaughtering. Before laws could be enacted, the indiscriminate killing of the guanaco decimated its population (estimated now to be less than 100,000), turning the guanaco into an endangered species. Yet poachers continue to hunt guanaco young, not because they remain a nuisance but because of their soft pelts. In contrast, the coat of a mature guanaco consists of a short, typically buff-colored undercoat that is overlaid with coarse dark-brown guard hairs. Not at all like vicuna hair, it also differs from the soft wool of the alpaca, wool that can grow up to 20 inches in length.

The alpaca, specifically bred during the third and fourth centuries B.C. for its shaggy wool, remains the primary source of raw clothing material in Peru. The color of most wool falls into one of 14 shades that range from brown to black; yet every so often, an alpaca with a white coat will appear, with wool that is prized almost as highly as that of the vicuna. And to produce even more desirable wool, the alpaca is sometimes bred with the vicuna, resulting in a long wool with a fine, light body. Of course the alpaca's long wool is not only of value to its breeder; it also serves to keep the animal warm in its harsh environment. The alpaca, raised in the alpine grasslands of central Peru (its principal residence) and western Bolivia, does best at altitudes between 13,000 and 15,000 feet, a bit higher than its presumed forefather, the guanaco, and just above the range of the llama.

Of all the South American camelids, it is the llama, with its banana-shaped ears, that gives taxonomists their biggest headache. Some contend that it is not a separate species at all but simply a domesticated guanaco that was bred to be a stockier and sturdier race. Others feel that while it may have derived from guanaco stock, the llama is now unique enough to merit status as a distinct species.

The llama, which lives in the Andes of central Peru, western Bolivia (home to 70 percent of all llamas), northern Chile, and northwestern Argentina, prefers the warmer grasslands and shrub country that lie between 7,500 and 13,000 feet. Its wool varies considerably among its widespread populations, some llamas bearing short, coarse hairs, others with longer and finer wool. Not a particularly valuable wool, it is used primarily to make ropes, rugs, and, only occasionally, clothing. Of much greater value is the llama's service as a pack animal. The male llama, domesticated by the Inca in the fifteenth century, is an ideal beast of burden. The largest and strongest of the lamoids, it can travel 20 miles a day over sparsely vegetated mountain terrain, carrying loads of up to 130 pounds. The females, always kept at home, provide their keepers with wool, fuel (from their droppings), and newborns.

Antelope ● Deer ● Pronghorn

Contrary to the song "Home on the Range," whose lyrics wistfully yearn for "a home where the buffalo roam and the deer and the antelope play," strictly speaking, neither buffalo nor antelope live in North America. What the song refers to as antelope are pronghorn, lone survivors of an old line of antelopelike animals that goes back at least 1.5 million years. Although antelope- and deerlike in appearance, pronghorn do not belong to the antelope family, Bovidae, or to the deer family, Cervidae, but constitute a family all their own, Antilocapridae. This family name, like the pronghorn's genus name, *Antilocapra,* is, unfortunately, a rather poor choice. Both terms were constructed from the Latin *anthlops* for "antelope" and *capra* for "goat"—the pronghorn is neither.

As an even-toed ungulate, the pronghorn, like deer, walks on the tips of its two middle toes, but whereas the deer retains a vestigial toe on both sides of its feet (often taken to be its lower legs), these appendages, called dewclaws, are absent on the pronghorn. Pronghorn also lack the lachrymal glands that deer carry under their eyes. But the pronghorn does have a gall bladder and specialized muscles for erecting the white hairs on its rump, features that deer lack (an exception is the musk deer, which does possess a gall bladder). Also, deer have no cartilaginous padding on the soles of their hooves, a cushioning device on pronghorn that enhances their footing and reduces shock to the legs when running. The pronghorn is also known for its keen eyesight, said to be comparable to a human using eight-power binoculars. This acuity is enhanced by bulging eyes (as large as those of a horse), giving the pronghorn an effective field of view of at least 180 degrees in each eye. Were it not for its unhealthy curiosity, the pronghorn would be difficult to take by surprise.

But its most distinguishing characteristic is the horns that are borne by both sexes. Although horns do not occur on all females, those that do tend to be more like spikes, rather than the longer, bifurcated designs carried by males. To the settlers of the American West, it was these horns that supported their contention that the slender, slim-legged pronghorn was closely related to the antelope of Africa and therefore merited the name "antelope." However, a closer examination of the pronghorn and its evolution eventually proved otherwise.

The horns of the pronghorn set these swift animals apart not only from deer but from almost all other ungulates as well—even those with horns. Unlike sheep, goats, cattle, giraffes, and all but one antelope (the Saiga of Russia), pronghorn shed their horns annually. Each year the pronghorn's horns begin anew from specialized "horn skin," which sprouts keratin (hairlike material) that forms a hard sheath. Constructed over a core of bone, the sheath reaches upward and then sharply hooks back and slightly inward to a point, but midway between its base and tip, the sheath also produces a short forward-facing spike or prong—the source of the animal's name.

Because of the Saiga and the few antelope that shed their horns once or twice during their life, some authorities feel that the pronghorn should remain classified in its old Bovidae subfamily, Antilocaprinae. However, others point out that the base of each horn's bony core most closely resembles that found in fossil mammals related to modern deer. The solution to these differences of opinion was the establishment of Antilocapridae.

Evolving into their present form during the Miocene period, pronghorn made their home among the open plains of North America, the only continent they have known. It is estimated that at one time, their population reached 40 million and was represented by at least 11 species. Dwindling to only 20,000 during the 1920s, the one remaining species *(Antilocapra americana)* has since rebuilt its population to nearly a million.

Standing only 3½ feet tall at the shoulder, the 4½-foot-long, 125-pound pronghorn buck is the fastest animal in the world—over a good distance (top speed of 50 mph for ¾ mile). The cheetah, while a faster sprinter (top speed of 65 mph for ⅓ mile), would fade long before the pronghorn began to tire. The pronghorn is able to achieve such speeds because of its remarkably large lungs, a large heart that pumps blood loaded with hemoglobin, and highly efficient muscles.

Whereas the swift pronghorn is safest roaming the plains, where predators have little to hide behind, the slower deer avoids open spaces if possible. Camouflaging woodlands, rocky terrain, and tall grass meadows all afford deer a much better defense against predators. Ranging in size from the 14-inch tall South American pudu to the 7-foot moose (shoulder heights), their most obvious feature is their display of antlers. Yet except for the caribou, only males sprout this bony and usually branched and cumbersome headgear. Antlers, commonly employed to mark territory, are also used for making threatening displays, sparring, and to impress does.

Yet antlers do not adorn all males. The musk deer and the Chinese water deer, both of Asia, instead bear long canine tusks. Because of this, most authorities do not classify these animals in Cervidae but in their own families, Moschidae (musk deer) and Hydropotidae (Chinese water deer). However, some zoologists do not embrace this classification, contending that the musk and the Chinese water deer better belong in Cervidae, next to the small muntjacs and the tufted deer—antlered deer that also bear tusks.

Because, according to some experts, antlers do not occur on all species, zoologists hesitate in using them—unique as they may be—to distinguish deer from other even-toed ungulates. Instead, scientists look to the evolutionary relationship between deer and their close relatives of Bovidae. This family of cattle, sheep, goats, and antelope consists of all the ruminating (cud-chewing), even-toed ungulates (hoofed feet) with hollow, unbranched horns. In general, the bovid body is much like that of deer: stout body, lanky legs, and long neck and snout. And despite the hesitation of some zoologists to cite cranial horns as a defining feature, such horns underscore the identity of Bovidae more than any other single feature.

Arising from a common ancestor, Bovidae and Cervidae went their separate ways during the early Oligocene period (37 million years ago). Those preferring the savannas and the grasslands characteristic of Africa adapted by developing features that today typify the bovids, whereas those favoring the cooler forests and woodlands of the north became more deerlike. Initially, the members of each group were not particularly diverse, but as they slowly radiated into more varied habitats, their adaptations gave rise to a variety of forms.

The deer line retained many of its original characteristics: long legs, medium-size body, long neck, short tail, scent glands, teeth with low crowns, and antlers. Variations in these features still easily distinguish the various species within the deer family: moose, caribou, elk, muntjac, pudu, roe deer, whitetail, and so on. But the bovid line, spreading into more diverse habitats, underwent an even more radical transformation.

Those invading cooler climates and hilly terrain became longhaired and short-legged creatures, the sheep and goats; others grew large and robust, the cattle; the rest, retaining a much slighter build, we call antelope. Of these three bovid forms, the antelope group is by far the most diverse. In fact, as Richard Estes observes in his book *The Behavior Guide to African Mammals,* "antelopes from different tribes are as unlike as are sheep and cattle."[2] Such extensive variation among species makes it difficult to define "antelope" any more precisely than the original gist of the word, which was "any bovid neither sheep nor cattle."

To early Europeans, who were quite familiar with deer, cattle, sheep, and goats, the few agile gazelle-like animals found living in Asia and Africa were something of a novelty. Bearing horns, rather than antlers, they were immediately recognized as even-toed ungulates related to sheep and cattle. The first of these to be described (in 1766) was the blackbuck of Pakistan and India. This lanky, spiral-horned animal was given the scientific label *Antilope cervicapra,* a name fashioned from the Latin *Antilope,* meaning "fabulous beast," and *cervicapra,* a combination of deer and goat designations. Subsequently, similar animals were viewed as types of antelopes, a notion that soon established "antelope" as an umbrella term for any bovid not fitting the sheep/cattle mold.

But as more and more such bovids were identified, the antelope group became uncomfortably complex. The delicate 6-pound, 8- to 11-inch-tall royal antelope was quite unlike the hulking 1-ton, 6-foot-tall giant eland. For zoology, the solution was to split the antelopes into nine subfamilies, each acquiring equal standing with the sheep and the cattle subfamilies. However, some zoologists felt that this put too much distance between the various antelope species and diminished the stature and distinction of the sheep and the cattle subfamilies.

One solution to the dispute, proposed by the American zoologist G. G. Simpson, was to introduce the intermediate rank of tribe to Bovidae taxonomy. Some antelope subfamilies he let stand, and the others became tribes within subfamilies. Unfortunately, neither this reordering nor any other using tribes has become widely accepted, leaving the relationships between the antelopes unresolved. This means that even today, a good definition of "antelope" does not deviate far from "any bovid other than sheep or cattle"—an ironic way of looking at them when one considers that they constitute the majority (84 out of 120) of all bovid species.

Although this is less than an ideal definition of "antelope," its rationale can be appreciated by looking at a typical Bovidae taxonomy (there are several), summarized as follows.

Subfamily:	Bovinae (wild cattle and spiral-horned antelope)
Tribe:	Boselaphini (four-horns)
Tribe:	Bovini (cattle; includes buffalo and bison)
Tribe:	Tragelaphini (spiral horns)
Subfamily:	Cephalophinae (small, fruit-eating, rainforest antelope)
Tribe:	Cephalophini (considerably arched backs)
Subfamily:	Hippotraginae (medium to large horselike antelope)
Tribe:	Reduncini (medium-size, wetland dwellers)
Tribe:	Alcephalini (large, high shoulders, long head)
Tribe:	Hippotragini (most horselike of all)
Tribe:	Aepycerotini (the impala, graceful with long neck)
Tribe:	Peleini (the gray rhebok, small and slender)

Subfamily: Antilopinae (small to medium-size antelope)
    Tribe: Neotragini (dwarf and delicate)
Subfamily: Caprinae (goats and sheep, mild–cold climate bovids)
    Tribe: Ovibonini (musk ox and takin, large and heavy)
    Tribe: Caprini (sheep and goats, smaller and stocky)
    Tribe: Rupicaprini (chamois and mountain goat, mountain dwelling)

Antler • Horn

Although horns and antlers can be exceptionally effective weapons, animals also employ them to mark territory, to intimidate rivals, and to attract the opposite sex. The only mammals that bear horns and antlers are herbivores—no carnivore is so equipped—so their primary use as a weapon, other than to duel with, is defensive, keeping predators at bay. Used in sparring bouts or as a threat, they aid in establishing rank within a herd. Although such headgear can serve as a formidable weapon, horned and antlered animals often opt to do battle with their hooves. This is commonly the case with deer, particularly after their antlers have been shed or before these have fully regrown. And even with a mature set of antlers, deer seldom use them to defend against predators. Instead, they will rear up and slash out with their forefeet. The pronghorn and the giraffe also find their hooves more effective in such situations (they do use their horns to spar among themselves); yet others, such as the rhinoceros and the buffalo, vigorously defend themselves with their horns. Animals seldom use their horns as tools, but the rhinoceros, faced with a shortage of traditional food (mostly grasses), has been known to use its horn to dig for roots.

Only deer carry antlers. Attached directly to the skull, these bony appendages are composed solely of bone. In contrast, horns occur on an assortment of animals and take several different forms. But whatever the form, almost all have one feature in common: a sheathed core of protruding bone. This is readily apparent in the horns of sheep, goats, cattle, giraffes, and pronghorn. For the narwhal (a marine mammal closely related to the beluga whale), however, whose "horn" lacks such rigid bony support, horn construction is considerably different.

The "horn" of the narwhal is a narrow, twisted shaft that juts out from above the mouth. Often called a "horn," the narwhal's strange spiraling shaft is actually a tusk—a tooth. Narwhal dentition is, at most, minimal; it consists of a single pair of 8-inch-long teeth set in the upper jaw. In females, tooth growth is conventional, but in males (infrequently females) one tooth (two teeth in rare cases) rears upward and, breaking through the upper lip, continues extending to a taper that can reach 10 feet in length, usually developing a left-hand twist as it grows.

Whereas the narwhal's spiraling tusk is not a true horn, the rhinoceros's odd weapon is. Its underlying bone is not prominent, like those of other horned mammals, but forms a modest knob on the rhino's skull (the maxilla), which offers support to a covering of skin. This covering arises from the same tissue that forms the sheathing of the bony core found in the horns of, for example, cattle. However, the rhino's skin does not transform into "horn," as is the case with cattle, but instead produces hairlike tubular fibers of keratin. Unlike true hair, which has a protective cuticle, these horn filaments are bare and more closely resemble the material found in hooves. A small amount of horn material firmly laminates these fibers together and molds them into a small spike that gradually enlarges into a hardened hornlike shape.

After a time, the fibers have a tendency to unravel and break off, but because, like hair, they continue to grow throughout life, the horn maintains its shape—abetted by sharpening against trees and other rough objects. Not directly affixed to the rhinoceros's nasal bone, the tenuously attached horn can be accidentally torn off or lost in battle. In either event, there is little bleeding, and the rhinoceros eventually regrows another. For the Indian and the Javan rhinoceroses that have but a single horn, such a temporary loss can be a definite liability, whereas those rhinoceroses with two horns, such as the African white, the Sumatran, and the African black (the latter sometimes sprouting a third horn) always have their backup horn—albeit usually the smaller one.

Among the remaining horned animals, the giraffe and the okapi have the simplest of horns. Growing in pairs, each horn is no more than a long, bony, skin-covered knob that emerges directly from the skull. They begin forming before birth as flat cores of cartilage, but within a week after birth each core swells and changes to bone. Each horn remains covered with hair until maturity, at which time the very top becomes calloused and turns black. In giraffes, a third, median "horn" may develop as a bump on the forehead, and in some races a second set will form behind the main pair. Late in life, a bull giraffe may grow other bony projections on its skull, giving the appearance of four or five horns. Typically, two of these develop over the eyes and one on the rear of the head.

The horns of the bovides (cattle, etc.) initially start out much like giraffe horns: simple bumps of skull bone. However, as these bony bumps grow, their covering of skin, instead of developing hair, hardens into "horn." This arises through the cornification of the epidermal cells, producing a covering of keratin—the same process that forms fingernails. Unlike its supporting bone, horn material is not supplied with blood. Although continuing to grow throughout life, like the rhinoceros horn, both the core of bone and its sheath of horn cannot be replaced if broken off. Because the bone and its covering of horn develop from different materials, their bond is somewhat tenuous, so after death the two structures often separate, leaving behind the familiar hollow horn.

Of all the head appendages carried by mammals, those of the bovides are by far the most varied. Antlers, with their various combinations of branches and tines, do assume a variety of configurations, but almost all follow a branching pattern (exceptions include those found on muntjacs, brocket deer, and pudu, which are no more than short spikes). The horns carried by the Cape buffalo are broad and massive, whereas those of the Japanese serow tend to be just tiny cones, which also contrast with the 3½-foot tapering spearlike set of the gemsbok and the curling horns that flank the head of the American bighorn sheep. With one exception, all horns, including those of the bovides, rhinoceroses, and giraffes (one could also throw in the narwhal), are unbranched—a decidedly striking distinction from conventional antler design. The single exception is the horn of the pronghorn antelope.

Not only are the pronghorn's horns noted for the short, forward-facing prongs that emerge about midway between their base and their backward-hooking tips, but the horns themselves are unusual in their construction. Like the bovid horn, each has a simple, unbranched bony core protected by a hard outer sheath. Because the hairs growing around the base of each horn tend to invade the sheath, and because the entire sheath has the texture of combed hair, the sheath was once thought to consist of consolidated hairs, similar to rhinoceros horn construction. But the pronghorn sheath is no more than cornified skin cells compressed into a hard shell and is very similar to bovid horn material. Unlike almost all other horns, a soft epidermal layer lies between the pronghorn sheath and its supporting bony core. It is from this membrane, which enrobes the bone, that the outer sheath develops.

Beginning at the tip of the bone, the sheath grows downward until it reaches the base of the horn and fuses with hair on the head. The sheath also extends upward from the bone tip, producing most of the horn's reach (the record length is 20 inches). During this growth, the sheath forms the prong that is characteristic of male horns. Between its base and the prong, the horn is compressed laterally, but from the prong to the tip it is cylindrical. The prong, which has a triangular shape that sometimes ends in a sharp point, is also flattened. Female pronghorn, which may not grow anything more than horn bumps, if that, seldom develop prongs.

In late fall, after the pronghorn's rut, the epidermal layer that lies between the bone and the sheath shrinks slightly and releases its grip on the sheath. At the same time, it also starts to form a new horn around the bony tip. Together, the two processes act to dislodge the old pronged sheath. Although this annual event is often mentioned as a crucial trait that supports the distinction between the pronghorn antelope and the bovides, such shedding of horn also occurs in species from all five of the bovid subfamilies, most notably among the Saiga antelope of Russia.

As with the pronghorn antelope, deer also shed their headgear annually. But unlike the pronghorn, when most deer shed their antlers, nothing is left protruding from the skull; the entire rack of antlers is lost. A short, flattened knob of skull bone from which the antlers emerge, called the pedicel, does remain, yet it is virtually invisible (exceptions occur among the muntjac deer, which have long pedicels, terminated by short, unbranched antlers).

Antlers, lacking sheaths of horn, are composed entirely of bone. They develop as extensions of the pedicels and retain coverings of skin, called velvet, until their growth is complete. The velvet protects and nourishes the bone, supplementing the bone's own blood supply. As antlers reach maturity, their bone begins to solidify, choking off its blood supply and, in effect, killing itself. A bit later, an expanding ring of bone growing at the juncture of the pedicel, called a coronet or burr, cuts off the blood supply to the velvet. No longer nourished, the velvet shrivels and dies. The deer rids its antlers of this dead skin by rubbing against trees and other objects, leaving the antlers no more than polished bone. The dead antlers remain attached to the living pedicel until testosterone levels fall below a critical level, at which time their attachment to the pedicel loosens and they are cast off. From its onset, this shedding process may take only a few days. The pedicel, left exposed, quickly scabs over—an essential development if there is to be antler regrowth the following year.

Antlers grow from their base at a rapid rate, a full set developing in only 12 to 16 weeks. From its appearance on a young animal to its full development on a mature buck, a set of antlers retains its basic configuration. Starting out as perhaps nothing more than a long spike of bone with a small branching, each successive pair of antlers builds on the previous year's design, becoming more complex as the deer ages. This continues until the deer reaches old age, after which the process goes into reverse. With each succeeding year, the aging buck produces antlers with fewer and fewer branches and tines, until its pedicels may bear only buttons of antlers—should the deer be fortunate enough to live so long.

Ape ● Monkey (Old World ● New World)

In the grand classification of life, apes and monkeys are placed in the order Primates—along with us humans. Comprising most of the order (147 of 183 species), they stand just behind humans as the most highly evolved of all the animals, many exhibiting a fair degree of intelligence. Excluding ourselves, the balance of the order (35 species) is made up of lemurs, lorises, aye-ayes, and pottos. These small, long-tailed tree dwellers, called prosimians ("pre-apes"), form the suborder Prosimii. All the other primates, humans included, make up the suborder Anthropoidea—a group markedly more advanced than the prosimians and therefore commonly called the "higher primates."

Of course, our anthropoid brethren the monkeys and apes sit far below us in the ranking of higher primates, with monkeys anchoring the bottom—a reflection of their early split from the main thrust of anthropoid evolution. As the anthropoid line continued to evolve, other distinct forms arose, a few surviving to the present. Among these, the earliest to have broken away were the gibbons, followed by the orangutans, and then the gorillas. Finally, around 5 million years ago the most significant split of all occurred, when the chimpanzees and the hominids (those leading to humans) went their separate ways.

It must be emphasized that primate evolution has not been linear. As with all evolutions, there have been many branchings—like a bush. Some branchings have had but a single representative; others have had many, and, of course, not all branches have made it into the present.

Zoologists informally dub orangutans, gorillas, and chimpanzees the "great apes," animals that comprise the family Pongidae. Those not so great, the "lesser apes," are the gibbons and their close relative the simang, which make up the family Hylobatidae. As for the monkeys, zoologists divide them into two distinct groups: Old World monkeys (Africa and Southeast Asia) and New World monkeys (Central and South America). Because of their differences, the two monkey groups were once thought to have arisen separately from the prosimians during the Eocene era (38 to 54 million years ago). However, recent biochemical assessments indicate it is more likely that the entire anthropoid line split from the prosimians at this time, and that it was not until the next epoch, the Oligocene (25 to 38 million years ago), that the two monkey lines branched off—the New World monkeys going first.

One of the features most often cited as distinguishing the two monkey branches is their noses: New World monkeys, with their wide-open nostrils that face outward and are set far apart, are formally called platyrrhines (broad-nose), whereas Old World monkeys, whose closed nostrils face downward and are set close together, are called catarrhines (downward-nose). The number of premolars they carry also differs: New World monkeys have a premolar formula of 3/3, whereas Old World monkeys have a 2/2 formula. In a New World monkey, the eardrum is supported by a single bony ring fused to the skull, whereas in an Old World species the eardrum is attached to a half-inch-long bony tube lined with skin. A prehensile tail, one that can grasp objects such as tree limbs, is also a mark of a New World monkey, although this ability is usually restricted to the larger species, such as the howler, the spider, and the wooly monkeys. No Old World monkey has a prehensile tail, but advanced thumb development does give its hands and its feet true grasping opposability, unlike the "pseudo-opposability" of the New World species. One can also identify Old World monkeys by the hairless calluses on their buttocks. Developing well before birth, these thick, cornified pads fuse directly to the ischial bones (the bottom bones of the pelvis) and serve as a cushion when sitting. As with humans, the nipples of an Old World primate sit on the front of the chest, whereas those of a New World monkey grow near the armpits, a site easily accessible to young monkeys who ride their mother's back.

Unlike the single family of Old World monkeys, Cercopithecidae, New World monkeys are divide into Cebidae (the capuchin-like species) and Callitrichidae (the marmosets and tamarins). The smaller, squirrel-like marmosets and tamarins, thought to be more advanced than the capuchins, each have a simpler uterus and only two, rather than three, molars (a single exception bearing three molars is the Goeldi's monkey, a tamarin). The prominent modern cusp pattern of their upper molars also points to a more recent development of the Callitrichidaes. But marmosets and tamarins sport claws, rather than the fingernails common to members of Cebidae. And whereas the two are avid tree dwellers, none have prehensile tails. Tamarins can be distinguished from marmosets by their longer limbs and by canines that exceed their incisors in length.

Like the marmosets and tamarins, the capuchin-like monkeys of Cebidae also prefer to live in the branches of trees, feeding on fruit, nuts, leaves, and small animals such as insects and lizards. This is similar to the food of the Old World monkeys, species that frequently use unique cheek pouches to store food.

Scrambling amid branches and swinging from limb to limb in search of a meal are typical monkey activities, particularly for those like the New World's spider monkey, whose prehensile tail complements its long limbs. Others, such as the more compact squirrel and the howler monkeys, prefer to leap and clamber, rather than swing. Most Old World monkeys also make their homes among tree branches; however, baboons, mandrills, the Patas monkey, and often macaques prefer living on the ground. To facilitate walking, the hands of these monkeys bend backward at the finger-palm joints, allowing the palms to remain vertical, which helps them to raise their heads and gives a better view. Whereas the fingers alone touch the ground, their feet remain plantigrade—the whole foot, not just the toes, contacts the ground. As ground dwellers, they have no real need for counterbalancing tails (as do arboreal species), so many have reduced or nonexistent tails. Lacking a tail, the Barbary macaque and the Sulawesi crested macaque have been regularly mistaken for apes and, as a result, are better known by their misnomers: Barbary ape and Celebes black ape. This, of course, repudiates the popular notion that the most conspicuous difference between monkeys and apes is the presence, or lack, of a tail.

The real differences separating the monkeys from the apes lie in the continuing evolution the anthropoid line underwent. The gibbons, as with many of the monkeys they followed, developed long, powerful arms, making them spectacularly adept at swinging among tree branches (an activity called brachiation). Relatively small and slight of build, male and female gibbons, as with monkeys, are similar in size, unlike all other apes who exhibit distinct sexual dimorphism. Their lightly convoluted brains and their buttock pads are also more monkeylike. Such characteristics reflect the gibbons' early split from the main line of anthropoid evolution and is why they fail to fit in with the "great apes" and have been assigned their own family.

Although monkeylike in many respects, gibbons are decidedly apes. As with the other apes, they have barrel-shaped chests (the monkey chest is more flattened from side to side), their forelimbs are longer than their hind limbs, and their wrist joints are constructed so as to allow a considerable range of motion. Unlike the cheek teeth of monkeys, whose grinding surfaces are made up of ridges, those of apes are crowned with conical bumps. Apes also have a more upright posture than do monkeys, and they lack tails. To be sure, the lack of a tail does not by itself distinguish apes from monkeys, but because almost every monkey has a tail, its absence on all apes makes it a helpful initial identifying mark.

After the gibbons, anthropoid development made a distinct break with the lesser apes. Succeeding apes grew much larger than the 12- to 23-pound gibbon: the average male chimpanzee weighing in at 130 pounds, the orangutan at 200 pounds, and the gorilla at 300 to 400 pounds. In all, the great apes are stout primates, with robust skulls, powerful jaws, and muscular forelimbs. However, as sexually dimorphic animals, the females show less development in these areas, always remaining the smaller and less robust of the two sexes. Yet like the males, they are in main still too big to deftly maneuver in the branches of trees, compelling the great apes to spend much of their time on the ground. But because their pelvic structures are not conducive to walking upright, most travel is done on all fours, the rear legs carrying most of their weight and, except for orangutans, the hands clenched for "knuckle walking" (gibbons are the only apes that are truly bipedal when walking).

The great apes also develop "sitting pads" on their buttocks, although these never become as prominent as those of gibbons, nor do they attach themselves to the ischial bones. Other characteristics of the ape, such as sparser body hair, a reduction of the cecum on the large intestine, greater finger dexterity, and very important, increased intelligence, all suggest that the anthropoids were continually steering closer and closer to a more humanlike form. In fact, molecular evidence indicates that chimpanzees share 97 percent of their DNA with humans, making them more closely related to us than they are to their nearest Pongidae relative, the gorilla.

Ass ● Burro ● Donkey ● Hinny ● Horse ● Jackass ● Jenny ● Mule

All ungulates are herbivores, whose evolution has produced a varied collection of animals that now numbers over 200 species. Using the number of toes on each foot as a basis of classification, science places them in two orders. By far the largest of these is Artiodactyla, the even-toed ungulates. It embraces nine families, including those of cattle, camels, deer, and pigs. The odd-toed ungulates, which make up the order Perissodactyla, are represented by only three families: the rhinoceroses, the tapirs, and the equids: the horses, the asses, and the zebras. The equids stand apart from the other Perissodactylas because their legs use only a single toe for support—the legs of rhinos and tapirs have three toes (each of the tapir's forelegs has an additional fourth toe).

"Equid" comes from an ancient Indo-European word for the horse and is used as a root for its family name, Equidae, and genus designation, *Equus*. Another ancient word for the horse, derived from Latin, is *caballus*, which united with *Equus* makes up the horse's scientific name: *Equus caballus*. The other members of the Equid family, all members of the genus *Equus*, are divided among eight species: *E. przewalskii*,[3] the wild horse of Central Asia; *E. zebra*, *E. burchelli*, and *E. grevyi*, the zebras; *E. kiang*, the kiang; *E. quagga*, the quagga; *E. hemionus;* the onager; and *E. asinus*, the African wild ass.[4]

The equids are an old line of ungulates that began their evolution some 70 to 60 million years ago in the forests of North America. Once small animals about the size of a fox, by 2 million B.C.E. they reached their present form and were migrating to the Asian continent. From Asia the equids eventually moved into Europe and Africa, where they flourished. For unknown reasons (perhaps the ice ages), those left behind in the Americas perished. Today, all the horses and the asses that roam the New World trace their ancestry to European stock brought over by early explorers and colonizers.

Although extensive breeding has distanced the modern horse from its predecessors, science has uncovered one species that is more closely related to the prehistoric equids than any other: the wild horse of Asia, or the Przewalski horse (pronounced sha-val-ski). Discovered in 1881, this small (four feet tall at the shoulder) horse was found living as a wild animal on the edges of the Gobi desert. It is distinguished by its large head, stocky body, short legs, and short, erect black mane. The Przewalski's coat is predominantly a buckskin color, while below the knees its legs turn dark and bear faint zebralike stripes. These stripes are also sometimes present on the shoulders and the back. Called a horse and appearing little different from several modern breeds, the Przewalski horse is genetically unique in the number of diploid chromosomes it carries, 66 (modern horses have 64; the African asses, 62; the Asiatic asses, 54–56; and the zebras, 32–46).

Although scientists are certain that the modern horse did not descend from the Przewalski horse, they have yet to pin down its true ancestry. What evidence there is points to a diverse genealogy consisting of several lines of evolution, which produced two principal forms of horse: Those from the warmer climates of Africa, Arabia, and Iran tend to be slender, be fast runners, and bear thin coats, whereas those from the colder climes of Europe and Northern Asia are stocky, are powerful, and usually possess heavier coats. These contrasting evolutionary features are not unique to horses but demonstrate Bergmann's Rule and Allen's Rule, two of three such rules that tie specific developments in a species to its climate.

But no matter how dissimilar the various breeds may appear, they all share several distinguishing characteristics. Besides their chromosome number, an unquestionably defining mark, it is their physical features that most obviously unite them. Foremost is their mane, which is typically long and never remains erect: it always lies to one side. The same is true of their forelock, an extension of the mane that grows over the forehead (some zebras also possess a forelock). And although the modern horse comes in a myriad of colors and patterns, no part of its body bears stripes.

Because of extensive breeding, certain characteristics once unique to other equines have since been bred into the modern horse, which has effectively reduced the apparent differences among species. A case in point is the Falabella of Argentina. The product of over 100 years of selective breeding, the Falabella is a shaggy, 3-foot-tall horse that on first encounter could easily be mistaken for a miniature donkey. Like the donkey, its mane is short, its body hair is coarse, and its rear legs are quite straight. However, the Falabella lacks the donkey's long ears, flat back, coarse-haired mane and tail, and white muzzle, white belly, and white eye rings. It also lacks the donkey's raspy Aw-ee, Aw-ee bray. These enumerated characteristics run true throughout almost all donkey breeds and go a long way in separating them from the horses. Other features distinguish the two as well. The donkey has a straight, rather than an arched, neck; a less than full-blown tail (its long tail hairs typically form only from the end of the tail bone); smaller and deeper hoofs; occasional leg stripes; and chestnuts (calluses) on the inside of only its forelegs; and it almost always lacks true withers (the shoulder hump at the base of the neck of horses).

These characteristics describe not only the donkey but also the ass. This is because "donkey" is no more than the common English word for the ass, *Equus asinus*. "Donkey" is said by some to come from two Old English words: *dun*, for the ass's grayish brown color (although asses do come in other colors), and *ky*, a diminutive termination indicating its small size. In the past, all donkeys were somewhat smallish animals that stood 30 to 36 inches high at the shoulder; however, through continued breeding there are now donkeys, such as the Poitou of France, that stand as much as 64 inches high—taller than the mustang horse.

Although donkeys are asses, the term "donkey" is usually reserved for those that have been domesticated, an endeavor that goes back at least 5,400 years. Zoologists trace their ancestry to the Nubian wild ass (thought to be extinct), although its striped legs also tend to tie it to the Somali wild ass—the only race of wild ass definitely known to still be around.

But the Somali's distinction as the last wild ass is only true if "ass" is restricted to the equids of *E. asinus*. Although the onager of Asia *(E. hemionus)* bears a few characteristics typical of horses (it is somewhat tall and its ears are not conspicuously long), its other distinguishing features are all asslike. About the only major feature that really identifies the onager is its number of diploid chromosomes, but whether this should prevent us from calling it an ass is open to argument.

One animal that is *definitely* an ass is the burro. Not a separate species or a breed, the burro is simply a donkey that lives in the western United States, Mexico, and other Latin American countries. We often associate *burro,* originally the Spanish word for donkey, with images of a small pack animal trudging along a dirt path or up a narrow street, its back laden with firewood or water jugs. Not surprisingly, this is a very valid image; other than occasionally being raced for amusement, the burro is primarily a beast of burden.

Usually standing no more than 30 to 34 inches high at the shoulder, the burro is one of the smaller donkeys and bears all the features typical of the donkey. Because it is sturdy, does well in arid climates, and is easily tamed, Spanish explorers brought it to the Americas as a pack animal. Still used in rural regions as work animals, many have escaped servitude and now make up small feral families living in the wilderness.

Reverting to the wild and creating feral families also occurs among horses and the other asses but not among mules. Although a male and a female mule might escape to the wild, they would be the last of their kind. Unlike the ass and the horse, species that easily perpetuate their lineage, the mule is not a true species, nor can it sustain a genealogical tree; it is a reproductive dead end, incapable of passing on its bloodline. (Although males are always sterile, a few females have produced young when bred with asses and stallions.) So, if mules do not beget mules, where do they come from?

They come from the mating of a horse and an ass. These two animals do this not by choice but through our prodding, a deliberate crossing of the species that goes back to at least 1200 B.C. And like most matings between two species, the result is almost always sterile offspring. This is true whether a mare (female horse) is mated to a jackass (a male ass), or a jenny (a female ass) is mated to a stallion (a male horse). In either case the hybrid foal will be a mule, although the offspring of a jenny is very different from that of a mare. In fact, their differences are so apparent that those born of a jenny are not normally called mules but hinnies—a distinction adopted for the balance of this discussion.

For reasons that lie within quirks of genetics, a mule tends to have the body of a horse and the extremities of an ass. It has also been observed that the rear of a mule is more horselike and the front end is more asslike. A mule is a good-size animal, often larger than either parent; it also has a full tail and well-developed hindquarters. However, as with asses, its legs are straight, its head is large but narrow, its mane is short, and its ears are long.

The hinny, in contrast, has almost opposite features. It is a smaller animal with a more rounded body; its tail is tufted like that of an ass, and its hindquarters are slighter. Like the horse, the hinny's legs are stronger; its head is smaller, as are its ears; and its mane is longer. In general, the mule looks more like a large ass, and the hinny, more like a small horse.

Although the mule and the hinny are both a mix of horse and ass characteristics, the mule possesses the qualities more widely prized by man. Used almost exclusively as a pack or a draft animal, it was given these jobs because of its large size and strength. It is highly regarded for its tough hide, resistance to disease, ability to withstand climate extremes, and ability to survive on native forage and little water. The mule is also very intelligent and easily trained, but wary and independent, which is often seen as stubbornness. The hinny, a mellower animal, is physically not quite up to the mule, yet it has not been totally ignored. Ancient records indicate that the Egyptians of old used hinnies to pull chariots, and that the Chinese had them hitched to carts. Today, jennies and stallions tend to be mated only when a mare and a jackass are not available, or when a novel pet is desired.

Bat • Rabbit • Rodent

Admittedly, this appears to be an odd threesome; do we really confuse bats or rabbits with rodents? Their differences seem all too obvious. Yet there are those today who do mistakenly believe rabbits to be rodents. A 1992 publication aimed at the outdoor sportsman began one statement with: "The varying hare of the north, like many other rodents . . ."[5] No less daunting is the old and persistent misperception that bats are rodents. One of the chief sources of this notion lies in the Chinese, the German, the Russian, and the French languages, where the word for "bat" translates as either "flying mouse" or "bald mouse"—recall Johann Strauss's operetta *Die Fledermaus*. While these "mouse" labels have since lost their credibility, they remain deceptive and reflect the confusion that still surrounds bats.

Fortunately, unlike *fledermaus,* "bat" is a distinct term with no confusing connotations. Yet the bats' resemblance to mice with wings nourishes the mistaken notion that this is truly what they are—winged rodents. And this is not a misperception held only by the uninformed; in the past, scholars were totally confounded by the bat's appearance and its relationship to other mammals.

Aristotle grouped bats with the rabbits and rodents. The sixteenth-century Swiss naturalist Konrad Gesner considered bats to be creatures fitting in somewhere between birds and mice. In 1693 John Ray put them with the armadillo and the hedgehog in Anomala, a special group he created for animals he could place nowhere else. Even Linnaeus, the father of taxonomy, had trouble with bats; he classified them with the primates: the monkeys and the apes. Today, bats reside in their own order, Chiroptera, a rank equal to those of the rodents (Rodentia), and the rabbits (Lagomorpha).

Without doubt, the bat's most distinguishing feature is its wings, making it the only mammal capable of flight. Other so-called flying mammals, such as the flying squirrel and the flying lemur (colugo), only glide: unlike bats, they cannot propel themselves by flapping. This unique ability of bats attests to a very successful evolutionary development; however, attempts to connect these mammals to an early ancestor or even to uncover a more recent progenitor have been unsuccessful. The oldest bat fossils, which go back to the Eocene era (about 50 million years ago), indicate that bats already had well-developed wings and looked pretty much as they do today. The best guess is that they most likely evolved 70 to 100 million years ago from shrewlike insectivores.

Rodents, however, are a more recent mammal and are thought to have arisen 55 million years ago in the late Paleocene. While science speculates that early rodents, like bats, were insectivores, nothing has been found to connect the two animals. So, given their long and separate evolution, it is not surprising that bats differ significantly from rodents.

Rodents, as do most mammals, have relatively typical forearms and fingers (the tarsiers being an exception), whereas bats have long forearms and remarkably elongated fingers. In fact, on most bats three of these fingers, which help form the wing's ribbing, are longer than the bat's body. The thumbs, however, are always quite short, some nearly enclosed by the wing membrane, others with small double claws or adhesive disks. To power their wings, bats evolved not only powerful upper arm and chest muscles but very large hearts as well. It is not uncommon for a bat to have a heart three times larger than that of a rodent of equal size. A rodent, lacking a need for a powerful upper body, displays more conventional muscle development, although it does have exceptionally large jaw muscles. Called masseters, these produce the strong forward and closing motion in the jaw, essential for gnawing through tough materials.

Tooth structure among rodents is fairly standard, although the total number of teeth does vary between 12 and 28. Each species has an upper and a lower pair of self-sharpening, chisel-shaped incisors, but none have canines or anterior premolars. Bat dentition, which ranges from 20 to 38 teeth, shows more variation. Some species use sharply pointed teeth for puncturing skin; others have flat teeth for grinding vegetation. Some have 2 premolars; others, none at all. Their canines range from small to quite large.

As is true for all animals, these dental characteristics reflect their diet, which for bats is quite diverse. Although best known as insect feeders, many species prefer fruit, nectar, pollen, fish, vertebrates, vertebrate blood, and carrion. Rodents, being primarily herbivores, dine on a menu limited more to bulbs, roots, grain, greens, nuts, and so on. Some do feed on small invertebrates such as insects and spiders, and a few others are carnivores.

Although bats have generally decent eyesight, most also have large ears, because they depend on their echolocating ability when flying at night (pinpointing an object by noting how it reflects a high-pitched signal). In some species these grow larger than the head itself, a trademark of the spotted bat. To help produce their high-pitched signal, bats in several families use grotesquely misshapen noses to precisely regulate their signal's frequency. Unlike the inconspicuous nose of the tomb bat, the nose of the flower-faced bat is an assortment of convoluted fleshy flaps that truly does look like a small bloom pasted on the bat's snout. No rodent nose is so peculiar. Noses among rodents are fairly standard, as are their ears, although the range of ear size within the order is notable. Mole rats have very minute ears, whereas those of the Abert squirrel are large and tufted.

Unlike the troubling similarities of basic body form shared by bats and rats, the distinctions between rabbits and rodents seem all too apparent. Neither much resembles the other—at least, not to the extent that the two would be confused. All rabbits and hares can be easily identified by three prominent features: ears that are at least twice as long as they are wide, well-developed hind legs that are longer than the fore limbs, and a naked Y-shaped groove that runs from the upper lip up to and around the sides of the nose.

Despite these features, for many years rabbits and their close relatives the hares and pikas were thought to be rodents. This conviction was based primarily on the similarity of their front teeth: chisel-shaped upper and lower incisors that continue to grow throughout life. In addition, each also has a large gap (diastema) in its dentition, extending from its incisors to its premolars. These dental characteristics make rabbits and rodents very adept at gnawing—"rodent" was formed from the Latin *rodere,* meaning "to gnaw."

Early classifiers, faced with these similarities, felt that rabbits and hares, along with the pika, formed a unique group of gnawing herbivores that made up a small but valid suborder of the rodent order. And so they remained until 1912, when zoologist J. W. Gridley recognized several striking differences between rabbits and rodents. Armed with this information and convinced that the rabbit's evolution was distinct from that of rodents, Gridley proposed that rabbits and their close kin be placed in their own order, which was subsequently designated as Lagomorpha (meaning "hare-shaped"). It was established that the unique similarities shared by lagomorphs and rodents were really the result of convergent evolution (comparable but separate adaptation), rather than a linked evolution. Fossils of even the oldest rabbits and mice fail to show any sequential evolutionary connection between the two.

The status of Lagomorpha, however, has not been without controversy. Some scientists suggest that it and Rodentia be combined under the cohort rank of Glires. Others dismiss this notion of close relationship and, in fact, see a similarity between the lagomorphs and members of Artiodactyla: the pigs, the camels, and the deer. Another suggests that they be placed with the elephant shrews in the grand order Anagalidia. And in 1996 genetic evidence indicated rabbits, in fact, may be more closely related to the primates than to rodents.

Despite these side issues, no one disputes Gridley's contention that certain characteristics of lagomorphs and rodents do indeed distinguish the two forms. Ironically, it is the very same anatomical feature that first united the two that now separates them: their teeth. Whereas lagomorphs and rodents share a dental anomaly—their ever-growing incisors—the rabbit has an additional set of peglike incisors in its upper jaw, located just behind the front chisel pair. The purpose of these rear incisors remains somewhat of a mystery, yet in earlier days when rabbits were thought to be rodents, this extra set of teeth was deemed significant enough to name the rabbit suborder of Rodentia, Duplicidentata (having duplicate dentition), and the suborder that included all other rodents Simplicidentata (having simple or basic dentition).

But aside from a second set of incisors, other dental features distinguish the two. Unlike the incisors of lagomorphs that have a jacket of hard enamel, those of rodents have enamel on only the front surface. Rodent gnawing, then, wears down the rear of the tooth faster than the front, producing a much sharper chisel edge than can be fashioned on rabbit incisors. Rats with well-developed and sharply honed incisors have even been known to gnaw through lead pipes and tin cans. Another difference in the dentition of the two is readily apparent if one watches a rabbit chew its food. Rabbits and hares will grind away at their dinner by working their lower jaws from side to side, not up and down as do rodents. This is because the two rows of molars in the rabbit's upper jaw are farther apart than are the bottom rows. So when chewing, the jaw must alternate between sides, bringing one side of upper and lower molars in opposition, then the other side. Rodents and lagomorphs also differ in the number of total teeth. Most rodents have 22, although the silvery mole rat has 6 extra, and the Australian water rat, 12 less teeth. Except for pikas and the Amami rabbit of Japan, both with 26 teeth, lagomorphs have 28 teeth.

Unlike the skulls of other mammals, a rabbit's skull is uniquely jointed. An encircling seam located toward the rear of the skull permits a slight movement between adjacent bones. The lagomorph skull is also distinguished from that of rodents by the numerous pores in its upper jawbone. Supporting the rabbit and the hare skulls are very flexible necks, enabling their heads to swivel 90 degrees, an act impossible for most rodents, who have short, stiff necks. But rodents have it over rabbits when it comes to dexterity. Because the forearm bones of the rabbit do not rotate inward like those of rodents, the rabbit cannot bring its paws to its mouth; therefore, whereas the common hamster can pick up a piece of lettuce to eat, the cottontail rabbit can only nibble on it from where it lies.

Oddly enough, the lagomorph has one feature found only among the marsupials and tree shrews; its scrotum lies in front of the penis, not behind. This has led people to mistake some males for females, thinking the small testes in the scrotum were teats. Lagomorphs also have a penis like that of humans; that is, it lacks a baculum, or penis bone. In rodents, as in many other mammals, the penis has a ridged shaft of fibroelastic tissue that is thrust forward to produce an erection.

Bison ● Buffalo (Cape ● Water)

If asked to name Africa's most dangerous animals, the first to come to mind would probably be the lion, the leopard, the rhinoceros, and perhaps the crocodile. It is unlikely that the African, or Cape, buffalo would make the list. Big game hunters and native Africans, however, consider this massive and powerful relative of the water buffalo to be one of the most deadly beasts on the continent. An agile creature for its size, it is reputed to have killed more big game hunters than any other animal, and, failing this, it has left many a victim badly maimed. Extremely aggressive when provoked, Cape buffalo charge with heads lowered in an attempt to impale or knock down their adversaries with their broad horns. If successful, the Cape buffalo may play with its victim, tossing it back and forth on its horns, or kneel on it and savagely tear it apart. But not all attacks result from provocation; old bulls have been reported to stalk their victims—in some cases, even humans. And if overpowered, the Cape buffalo will fight to its death.

A truly savage animal, it has learned well to take advantage of its deadly weaponry: horns that emerge close together from the top of the head and that are connected by a broad bony "boss" lying across the brow. From the boss, its horns fall to the sides of the head over the ears and then recurve upward. These horns, growing largest among the males, can reach up to 55 inches in length. A smaller subspecies, the forest buffalo, carries shorter, well-spaced horns that curve back, rather than downward.

In contrast, the water buffalo of Southeast Asia is a more docile beast. This domesticated form of the Asiatic, or Indian, buffalo—an animal just as formidable as the Cape buffalo—has been in service to man for over 4,000 years. Enlisted in many countries as a source of milk, water buffalo also serve as draft animals for plowing agricultural fields— typically, water-laden plots of land such as rice paddies.

The water buffalo, a larger animal than the Cape buffalo (2,300 lbs. versus 1,700 lbs.), stands 6 to 7 feet tall, which is 10 to 15 inches taller than its African counterpart; however, its total length is no greater. Its broad horns grow from the sides of the head and sweep in a wide arc to the rear. These reach their greatest size in the undomesticated females of the wild, some of whose horns measure over 6 feet in length. All of the water buffalo's close relatives, such as the tamarau and the anoa of Indonesia, are smaller species, with horns typically between 12 and 20 inches in length.

Although the water buffalo and the Cape buffalo exhibit significant differences—they split from a common ancestor 3.5 to 4.5 million years ago—as buffalo, they share stout bodies; low, wide heads with broad snouts; large ears; relatively short necks; long tails ending in tufts; and splayed hooves, but none have the facial or foot glands characteristic of others in their family, Bovidae.

As with all bovines, both easily become dehydrated; therefore, a reliable water source is crucial for survival, particularly during the arduous dry season. When the rains do arrive, these animals make up for their hardship by spending hours wallowing in mud. In fact, water buffalo favor watery terrain, preferring swamps, marshes, rivers, lagoons, and humid rain forests. The Cape buffalo, for whom wetlands are a scarce resource, is more often found in the savannas or the high grass meadows of wooded country but never straying too far from water.

The bison of North America, aside from enjoying a more temperate climate, inhabits very similar territory. The plains bison, one of two subspecies, favors open prairies, whereas the larger and darker woods bison is more a forest dweller. Evolving fairly recently from a common ancestor, the two once roamed the greater part of the North American continent. During the seventeenth century, the estimated 60 million bison of North America extended from the Canadian southwest, down to and across almost the whole of the United States and into the northern half of Mexico. Today, the surviving bison population of 50,000 is split among 19 major herds that are confined to state and federal parks, national reserves and refuges, and private ranches. Its European relative, the wisent, is also well protected in breeding reserves, where its population of 1,000 is steadily recovering from a near extinction low in 1922 of just 66 individuals.

Most often called a "buffalo" in North America, this alternative name for the bison arose as one of several variations of the English "la buff," a term taken from the French word for oxen, *les boeufs*. But this large, shaggy member of the cattle tribe is far from being a true buffalo. Neither a member of the water buffalo genus *Bubalus* nor of the African buffalo genus *Synceros,* the two bison species, the American and the European, make up their own genus, *Bison.* "Bison" is also the term science favors as their common name, a preference adopted to distinguish the animal from true buffalo.

The term "bison" is actually the Latinized spelling of the English "wisent." "Wisent," borrowed from the German language and now used to denote the European bison *(Bison bonasus),* originally identified the now-extinct auroch, an oxen. For the two subforms of North American bison *(Bison bison),* classification has not been as well established. Some authorities had regarded the plains buffalo as a species and woods buffalo as a race of plains buffalo, but today both are considered equal subspecies, with the plains buffalo classified as *B. b. bison* and the woods buffalo as *B. b. athabascae.*

Although bison belong to the same tribe (Bovini, the wild cattle) as do buffalo, fossil evidence pushes their common ancestor back 5 to 6 million years. Much more closely related to bison are the true cattle of the genus *Bos,* the genus of humped cattle, yaks, and all domestic cattle breeds. In fact, the two are so close that they easily crossbreed, an undertaking now promoted to produce "cattalo" or "beefalo," animals destined for the butcher market. Because the offspring of a male bison and a female cow are usually too broad shouldered for the cow to deliver safely, female bison are used in breeding. Sterility among male beefalo, once a frustrating problem, has now been fairly well solved by crossbreeding.

Even though bison and cattle are closely related, it appears that their lines split over 2.5 million years ago during their residence on the European continent. Bison, finding their way into North America by means of the Bering Strait land bridge during the ice ages, eventually spread all the way into South America. Although adapting to their new environment and ultimately evolving into the species we know today, they remain very similar to the European wisent.

Like buffalo, some bison do browse (eat the tender branches, twigs, and shoots of trees and shrubs); however, most prefer to graze on grasses. Slowly roaming the countryside in search of food, the bison is an unhurried, gregarious creature with a mild disposition, fostered, no doubt, by a lack of real enemies—except for man. When approached by man, a bison will just as often amble off as, taken with curiosity, stand its ground, perhaps growling or grunting at the intruder—true buffalo never growl but make more of a lowing sound or an occasional grunt. As with the Indian buffalo, bison can be tamed, but with little reason to do so, this has seldom been done. Although not easily excited or particularly dangerous, like many threatened animals these large beasts become aggressive when angered. Emulating the buffalo's savage attack, a bison will use its horns to impale its victim, throwing it into the air and then trampling it to death.

This is not difficult for an animal that stands 6 feet high at the shoulder and weighs between 1,400 and 2,300 pounds (American bison). The wisent, weighing a bit less, is the more slender of the two and is also distinguished by a shorter coat covering its forequarters, longer legs, a smaller head, and slimmer horns. Wisents also have a much smaller hump than do American bisons. This distinct feature, appearing to be formed by a spinal column that gently rises from the haunches to an apex above the shoulders, is really quite conventional: bison spinal alignment is no different from that of any other animal in its tribe. The bison shoulder hump is actually a distortion of the thoracic vertebrae, in which successive

spinous processes (blades emerging from the top of the vertebrae) gradually lengthen—the longest averaging 15 inches. This exaggerated growth
abruptly stops at the shoulder, leaving the head to appear lowered and
more aggressively positioned.

Its massive forequarters and head also set the bison apart from the
other wild cattle of Bovini. Especially striking is the large mass of muscle
lying across its shoulders. Enhancing this robust appearance is its heavy,
darkly colored coat—particularly during the colder months. In the spring
the bison sheds the heavier layer of its coat, leaving the remaining hide to
turn several shades lighter. In contrast, the buffalo, which lives in
warmer climates, has a very short and sparse coat, which on the water
buffalo, oddly enough, grows forward up the middle of the back. The
broad head of the bison, big as it is, appears even larger because of the
thick beard of hair attached to its dewlap. This loose fold of skin dangling
from the throat tends to remain relatively hairless in buffalo species.

But the feature that taxonomists see as really distinguishing the two is
the 14 pairs of ribs carried by the bison and the 13 pairs of the buffalo.
The horns of the bison also set this animal apart from the buffalo. Widely
spaced and growing from the sides of the head, bison horns tend to be
very short, some never reaching much more than a foot in length. Moreover, unlike the triangular cross-section typical of buffalo horns, that of
the bison horn is circular.

Bobcat • Lynx • Wildcat

The lynx, despite its exotic-sounding name, is a relatively unexceptional
cat whose name merely alludes to the way its eyes shine in the dark.
Coming from the Greek *lúgx* by way of Latin, "lynx" means "seeing,"
"illuminating," or "shining" and in this case refers to the ability of the
lynx's eyes to reflect light, a characteristic common to all cats.

This small cat, weighing 10 to 60 pounds and standing no more than
30 inches high, is better distinguished by its exclusive representation of the
cat family in the more northerly regions of Asia, Europe, and North America. The Canadian lynx, *Lynx canadensis,* and European lynx, *L. lynx,*
make their home among the coniferous forests and the scrub of the northern and montane regions. They also sometimes venture into the arctic tundra, but only if there is significant vegetation—rarely do they roam open
meadows. The lynx's preferred haunt is a dense forest, which affords it
camouflage in its hunt for small prey. For the lynx of Europe and Asia,

such prey typically includes rodents, rabbits, hares, ground birds, fish, foxes, and even small deer in winter. The diet of the Canadian lynx is similar but is focused more on the snowshoe hare. In fact, the existence of the Canadian lynx is so dependent on this little creature that its population rises and falls with corresponding changes in the snowshoe's population.

The Spanish lynx, *L. pardellus,* thought by some to be an offshoot of the European lynx, is a smaller (24 pounds) version that lives in central and southern Iberia. Like its northern relatives, it, too, lives off small game, plus an occasional deer. Today, however, urbanization, deforestation, and a plunge in the rabbit population because of myxomatosis (a virus) have left the Spanish lynx scrambling for survival. With a Spanish lynx population reduced to between 1,000 and 1,500 in the 1970s, conservationists began introducing it into Yugoslavia and Austria in an effort to save the animal.

All lynxes are united by two distinctive features. Most striking are their ears, which are crowned with two-inch spikes of glossy black fur. Scientists speculate that these tufts may serve to improve hearing or their movement may be used to signal other lynxes. The other unique feature of the lynx is its large furry paws, which act like snowshoes, allowing it to walk on soft snow. In main, the lynx is a short-bodied cat with strong limbs and a black-tipped, bobbed tail. Its coat varies from a smoky tan to pale rust and is relatively unmarked among those living in North America. The coats of European and Spanish lynxes, however, develop prominent spots, particularly in summer when the pelage is short. The winter coat of all the lynxes is noticeably long and grows densest among those of North America. The lynx also sports an exaggerated growth of fur on its cheeks, called a ruff, which in North America has often led to its misidentification as a bobcat.

The bobcat, *L. rufus,* a strictly North American cat that sports a facial ruff, might be confused with the lynx only where their territories overlap, in a broad region that runs across southern Canada. The balance of the bobcat's range extends into Mexico and covers all but the upper eastern half of the United States.

Because this is such an extensive, climatically varied realm, the bobcat will be found not only in dense forests but in brushland, semideserts, swamps, and canebreaks as well—it does not do well in deep snow. This, in turn, affords the bobcat a much broader menu than that of the lynx. Rodents and the cottontail rabbit, in particular, are a favorite, but fish, poultry, frogs, snakes, prairie dogs, domestic sheep, pronghorn—in fact, anything it can catch—help round out its diet.

In winter, the bobcat's grayish coat is long and dense but not to the extent of the lynx's coat. In summer, the coat is much shorter and considerably redder, and its mottled spots that were muted in winter become conspicuous. Although the bobcat also sports token ear tufts, this cat is readily distinguished from the lynx by the black-rimmed white splotches on its rear and its bobbed tail (the source of its name), which is proportionally longer than the lynx's, has indistinct black stripes on top, and is white underneath. Both of these markings are thought to help kittens follow their mothers at night. In all, the bobcat is a smaller cat than the lynx and is not as streamlined. It also lacks the lynx's broad paws, and its limbs tend to be shorter and more slender.

Frequently called a "wild cat" and therefore mistakenly thought to be rather savage, the bobcat is not a particularly fierce cat at all—no more so than any other undomesticated cat. "Wildcat" is actually a misnomer, a name stolen from another small cat, the European wildcat, one of the most ferocious cats in the world.

With a body averaging two feet in length and weighing no more than 15 pounds, the wildcat will not hesitate to attack an intruder many times its size, humans included. This aggressive disposition, however, is well masked by its benign appearance. Pictured alongside the striped domestic "tabby cat," the two are virtually indistinguishable, although the wildcat will probably be somewhat larger and have a blunted, black-ringed tail. This resemblance is not surprising, inasmuch as all domestic cats are descendants of ancient wild cats that are closely, if not directly, related to the European wildcat. The most likely ancestor of the domestic cat is the African wildcat, although the slightly larger European wildcat may also have been involved. In some cases, those European wildcats found living near human populations are crossbreeds between genuine wildcats and feral domestic breeds. But unlike the domestic cat, which readily takes to human companionship, the wildcat refuses to be tamed, insisting instead on living up to its name.

Like the bobcat, the European wildcat lives primarily off small mammals such as rodents and hares, as well as an occasional grouse. Wildcats in Scotland live principally in the highlands, hunting among whatever brush is available. Likewise, those in Asia Minor, the Balkans, and Central Europe prefer living in forests or at least amid dense scrubland. These three wildcats are so closely related that their previous designations as separate species have now been downgraded to subspecies rank. Classified in *Felis silvestris*, the European wildcat is listed as *F. s. silvestris*, the African wildcat (also called Caffer cat) as *F. s. lybica,* and all domestic cats as *F. s. catus*.

Brown Bear ● Grizzly ● Kodiak

Except for the polar bear, no other terrestrial carnivore in the world approaches the size of North America's brown bear, *Ursus arctos*. The average weight of these bears (depending on specific population) ranges between 250 and 1,000 pounds, with the record holder, a Kodiak, weighing in at 1,656 pounds. The disparity in average size is principally a result of available food. Those living near coastal waters where fish is abundant typically weigh 300 to 500 pounds more than their inland counterparts, bears whose diet consists mostly of vegetation. Their classification as carnivores, meat eaters, does not mean they eat no other foods; all bears are omnivorous.

In North America it has become a common practice to use the term "grizzly" for those of *U. arctos* living inland and "brown bear" for those living on the coast. However, this distinction, instigated by hunters, is altogether artificial. In the early 1960s the Boone and Crockett Club, the official record-keeper for game taken in North America, was pressured by bear hunters into designating those of *U. arctos* living within 85 miles of the coast as brown bears and those living farther inland as grizzlies. This recognition of two distinct forms was concocted to give the hunter who was looking to bag a trophy inland bear a "sporting" chance against those hunters who pursued the larger coastal bear. Although the distinction may serve the egos of hunters, it has only added to the confusion involving the two bears. In fact, the distinction has become so entrenched that some wildlife "authorities" have come to regard "grizzly" as the official designation for *U. arctos*. They not only use "grizzly" as the species' common name but call any brown bear living near the coast a "brown-bear grizzly." In truth, it's just the opposite that better fits the facts: the grizzly is but one of several forms of brown bear.

Of the eight bear species that make up the family Ursidae,[6] brown bears have, by far, the widest distribution. Inhabiting not only North America but northern Japan, northern Asia, and Europe, they exist in an assortment of sizes and colors. The reigning king of the brown bears, the Kodiak, which lives on the Alaskan islands of Kodiak, Afognak, and Shuyak, may weigh 1,000 pounds or more and stand over 10 feet tall. The smallest brown bears, typically those of Western Europe, rarely reach 600 pounds and seldom stand more than 7 feet tall. Despite the brown bear's name, its coat ranges from cream to various shades of brown to almost black. But the true grizzly, whose name has been so sorely misappropriated, was in fact named for the color of its coat.

"Grizzly" is generally regarded as coming from the French *gris* and Middle English *grisel,* both meaning "gray."[7] It referred to the somewhat "grizzled" appearance of the animal's coat (mainly on its back and neck), which is produced by the clear and lighter-colored tips of its guard hair. The coat is not a true gray but rather brown with a blond to silvery cast to it—one of the grizzly's nicknames is "silvertip." Where the tinting covers almost the entire length of the guard hair, very little of the dominant brown remains, and the animal will appear cream or tan.

Grizzly bears are all inland brown bears, but not all inland brown bears are grizzled; in fact, brown bear populations throughout the world exhibit a tremendous variation in coloration. Some inland groups have an almost completely pale-yellow back, some carry only a tinge of silver, and others show no hint of grizzling whatsoever, all of which makes "grizzly," despite its wide recognition as a valid name (even as a synonym of "brown bear"), a poor, if not misleading, label for inland-dwelling brown bears.

But hunters are not the only ones responsible for the confusion surrounding the brown bear and its various names. Naturalists and taxonomists, in their ongoing struggle to categorize the various forms of brown bears, unwittingly supported the excessive number of common names once laid upon these animals. When Carolus Linnaeus first classified bears in 1758, he designated them *Ursus arctos: Ursus* from Latin and *arctos* from Greek, both meaning "bear." In 1815 the naturalist George Ord refined this classification by placing the polar bear in *U. maritimus,* the black bear in *U. americanus,* and the grizzly brown bear in *U. horribilis* (meaning "horrible bear"). Other suggestions for the "grizzly" made around the same time included *U. ferox, U. candescens,* and *U. cinereus.* The Kodiak, initially not recognized as a separate form of brown bear, was eventually classified as *U. middendorffi* in 1896.

This recognition of additional species is indicative of the trend in taxonomic classification around the turn of the century that focused on differences and tended to separate various forms of an animal. By 1918, 86 species of "grizzlies" had been identified, a formulation devised by the biologist Dr. C. Hart Merriam. The remaining brown bears of *U. arctos* underwent a similar splitting, leaving us with, among others, *U. a. ardsoni* (barren ground "grizzly"), *U. a. gyas* (peninsula brown bear), and *U. a. utahensis* (brown bear of Utah).

This liberal classification and variations of it persisted until 1962, when Dr. Robert L. Rausch, focusing on the similarities rather than the differences of brown bear populations, performed radical surgery on their taxonomy. By carefully comparing skull features and weighing other factors, he concluded that the various populations of distinct subspecies were simply communities of brown bears that represented a broad integration of characteristics. Dr. Rausch also concluded that all brown bears (grizzlies and Kodiaks included) belonged to but one species, *Ursus arctos;* and of those in North America, only two true subspecies could be identified. The Kodiak, because of the shape of its claws, its unusually massive skull features, and its dentition, was separated from the others and classified as *U. a. middendorffi.* All other North American brown bears were assigned to *U. a. horribilis.* Not as clear is whether those outside North America, such as the Eurasian or Western European brown bear *(U. a. arctos)* and the Hokkaido brown bear *(U. a. yesoensis),* should retain their separate subspecies designations. Not all taxonomists subscribe to this extreme lumping of the North American brown bear into two subspecies, but because a more persuasive classification has yet to appear, most authorities defer to Dr. Rausch's assessment.

The main message of this reordering is that there are no specific animals such as the grizzly bear, the Alaskan brown bear, the Chinta bear, the Dall brown bear, the patriarchal bear, the peninsula brown bear, the tundra bear, or any of the other so-called species and subspecies of brown bear. These bears, some now extinct, only represent variations inherent in *U. a. horribilis.* The true grizzly is simply a brown bear with a tinted coat. And as with the "85-mile inland limit" established by the Boone and Crockett Club, deciding the degree of tinting necessary to make a grizzly out of a brown bear is entirely subjective.

The grizzly may be smaller and more irritable, adversarial, and unpredictable than other brown bears, but as its population approaches coastal regions, these characteristics, along with its coloration, diminish and eventually disappear—there is no abrupt change. Even the Kodiak, a distinct subspecies, is not far removed from its coastal relatives. As Harry V. Reynolds of the Alaska Department of Fish and Game observed in 1987: "There appears to be little difference between the size, population biology or behavior of bears on Kodiak Island and the nearby mainland bear population."[8]

Because of their reclassification, the old confusing references to various brown bears may eventually disappear, although "grizzly"—stripped of its original meaning—seems destined to become their standard name among North Americans. But no longer should we read that the Kodiak is a "strain of grizzlies"[9] or a race of Alaskan brown bear (the Alaskan also having been identified as *U. a. gyas* and *U. a. middendorffi*), that the coastal brown bear is but a variation of grizzly, or that grizzlies can be found throughout the world. Of course, that we should no longer encounter such misleading declarations certainly doesn't mean that we won't. On the Discovery Channel's *Wild Discovery* program of September 30, 1997, "The Giant Grizzlies," the Kodiak was referred to as a "Kodiak grizzly." Not only have the brown bear's old labels become ingrained in the public's mind, but many people (including wildlife writers) appear ignorant of the implications associated with the brown bear's updated taxonomic status—if they are aware of it at all.

Camel ● Dromedary

The dromedary is a camel—not a camel-like animal, not a close relative of camels, but a bona fide camel. Any impression to the contrary may well have been gleaned from the conflicting information presented in various reference books—more than one source has failed to clearly establish this simple fact. For example, from a noted encyclopedia on animal life, "Camels *and* dromedaries have become accustomed to kneeling down."[10] From another otherwise excellent book on animals, "Camels are the largest of the artiodactyls. The legs are long (longer in dromedaries *than* camels)," and "The gestation period of a camel is about 13 months, *and* slightly less, about 12 months, for the dromedary."[11] (All emphases added.) The text accompanying these quotes indicates that the authors are aware that dromedaries are camels, but the reader may well be left guessing: dromedaries = camels?

Camels come in two basic forms: the two-humped Bactrian camel *(Camelus bactrianus)* that lives among the steppe grasslands and the deserts of Central Asia and northern China and the one-humped Arabian camel *(Camelus dromedarius)* that makes its home in the deserts of North Africa and Southwest Asia. Their species names are clues as to which is better known as a dromedary, but only those Arabian camels bred for riding and racing truly merit the name "dromedary." Unlike the common Arabian "baggage" camel, the "true" dromedary is a lighter and lankier animal, particularly the North African Mehari dromedary, which was specifically bred for racing (top speed of 18 mph). "Dromedary" and the species designation *dromedarius* in fact derive by way of Latin from the Greek *dromas,* for "runner" (akin to the source of the suffix "drome," as in the words "hippodrome" and "aerodrome"). The term "Bactrian" comes from Bactria, an ancient area now a part of northeastern Afghanistan.

The ancestor of today's Arabian camel, which was an Asiatic animal closely related to the Bactrian, adapted to North Africa's hot, windy, and sandy environment by modifying several features. It evolved broad hooves to keep from sinking into the sand (although this is disputed by some authorities, who feel that broad hooves developed in Asian species in response to the deep snow they encountered), and its lips became tough and thickened in order to dine on coarse vegetation. Also, as protection from blowing sand, it developed small hairy ears, a double row of interlocking eyelashes to shield its eyes, and the ability to close its nostrils. Able to drink up to 30 gallons of water at a single filling, the Arabian can survive up to 3 weeks before needing a drink. Although its strength will wane if such a period grows too long, the Arabian camel can lose up to one-quarter of its body weight before its life is endangered. To help conserve water, it excretes very little urine and only dry feces; and to limit water loss through sweating, it delays cooling itself until its body temperature has risen a good eleven degrees. Its hump is also thought to aid in desert survival by serving not only as a stored food reserve of fat but also as insulation against the hot sun. Although the hump stores no water per se—as is sometimes believed—when its fat reserve is broken down, the released hydrogen combines with oxygen to form water within the animal's tissue.

Camels, having come from a chilly North America about 2 to 3 million years ago, arrived in Asia with a heavy coat, which still serves the Bactrian camel well during the cold of winter. This dense winter coat consists of 10-inch-long, ruddy-brown, wooly hairs. These grow thickest on its hump, upper forelegs, tip of tail, and particularly the neck, where its dangling mane may reach 2 feet. At the approach of summer it sheds most of this long coat, leaving behind only the lighter-colored short hairs. This dramatic change in its coat allows the Bactrian to withstand temperatures ranging from −15° F to 122° F. In contrast, the beige-colored pelage of the Arabian camel consists of fine, soft, wooly hairs that remain short throughout the year.

Because of its shorter legs, the Bactrian is about half a foot shorter than the 7-foot-tall Arabian camel (shoulder height). Its feet are also shorter and are protected from rocky terrain by hard, calloused soles. These features enable the Bactrian to easily negotiate barren hill country that the Arabian would find difficult at best.

Despite these contrasting features, which seemingly point to two very distinct species, some anatomists feel that such features may be no more than adaptive modifications in a species that better define two geographical races, rather than separate species. They contend that the difference in leg length is simply a minor variation; that whereas the Bactrian camel does have a second hump, the Arabian camel in its embryonic form also shows a vestigial second hump; and that interbreeding (as done in Turkey with a Bactrian male and an Arabian female), which produces fertile one-humped females, is indicative of a single species (male offspring are usually sterile).

While their status as separate species may not be resolved to everyone's satisfaction, the Bactrian and the Arabian do remain two distinct forms. Distinguished not only by their appearance, they also have slightly different dispositions, which influenced their domestication. The Bactrian camel, first tamed in the Mideast around 2500 B.C., was eventually brought to India and China, where its mild temperament, coupled with its physical strength and endurance, made it the favored beast of burden. For centuries, these camels diligently transported nomadic tribes and their households across central Asia and also made up the merchandise-laden caravans of traders. As a domestic animal, the Bactrian camel also serves as a source of milk, meat, wool, and leather. Unlike the Arabian camel, which exists only as a domesticated animal—or, as in Australia, a feral renegade—there are still 300 to 500 wild Bactrians living in the Gobi desert near the borders of China and Mongolia that have been able to perpetuate an unbroken lineage of wild ancestors.

The Arabian camel serves much the same needs as the Bactrian does, although its servitude may go back no more than 14 centuries (some authorities contend that it was first domesticated as long ago as 4000 B.C.). Inherently more irascible than the Bactrian camel, Arabian camels tend to be obstinate, malicious, and even "stupid" beasts, making the Arabian "baggage" camel a somewhat troublesome beast of burden. But even more of a disciplinary problem are the racing dromedaries, the most highly strung and temperamental of all camels.

Caribou ● Reindeer

Of all the deer in the world, from the lumbering moose of the north to the 14-inch-tall southern pudu, only among the caribou and the reindeer do both sexes bear antlers, though those of the female are typically smaller and have fewer points. Antlered females occasionally do occur in roe, whitetail, blacktail, and mule deer, but such anomalies are usually infertile.

Although antlers are a conspicuous feature among female caribou and reindeer, they do not occur on all does. In the Ungara caribou population of Newfoundland, as many as 30 percent of the females lack antlers. Because antlers usually serve only male needs (combat displays, mating lures, territory markers, ranking mechanisms, etc.), their appearance on females is a bit odd. Compounding this peculiarity is the fact that does retain their antlers much later into spring than do bucks. It has been suggested they do so to compete with bucks for forage during their pregnancy, but neither this notion nor other attempts to answer the mystery of doe antlers have proven satisfactory.

Caribou and reindeer, almost identical animals that dominate the northern tundra and bordering woodlands, are considered to be a single species, *Rangifer tarandus.* However, this has only been true since the 1950s. Previously, because of their size, coloring, and local groupings, caribou were broken down into several species, *R. arcticus, R. caribou,* and so on. But when taxonomists decided to change the emphasis in caribou classification from caribou differences to their similarities, the various species were downgraded to subspecies status and placed in the reindeer species taxon, *tarandus* (some authorities regard them as races rather than subspecies).

The word "caribou" is an Algonquin term meaning "pawer" or "scratcher," which describes the caribou's method of finding food (primarily lichens, grasses, and fungi) on the barren stretches of tundra. It is a term used only for those found in North America and sometimes only by North Americans, who also restrict the term "reindeer" to those subspecies living in Eurasia. In Europe, "reindeer" (coming from the Old Norse *hreinn*) is also the favored name for those of the Old World and often the North American caribou as well. But whatever one's preferred label, the two animals do display readily apparent differences, as do individual groups of caribou.

Probably the best-known caribou are the barren ground subspecies *R. t. granti* and *R. t. groenlandicus,* which range across the Yukon, British Columbia, Alaska, and Greenland. Often photographed making their seasonal migrations, countless bands of 5 to 100 have been seen merging into herds containing as many as 300,000. The barren ground caribou tends to have a pale brown to white pelage with a bluish-gray back saddle, which it retains throughout the year. The woodland caribou, *R. t. caribou,* is a darker subspecies, with a summer coat of short, dark sepia fur that covers all but its white belly and rump. In winter, this coat turns long and silky and takes on a grayish coloring. Because it feeds on forest vegetation, the woodland caribou is a bit heavier than its barren ground brother (400 lbs. versus 375 lbs.).

Their antlers also distinguish the two forms. The splayed rack of the barren ground subspecies tends to be slender and have shorter tines. The heavier antlers of the woodland caribou grow more upright (to better maneuver among trees and brush) and have more pronounced palmate ends. This subspecies, once common to the coniferous forests across the northern United States, now numbers only around 43,000 and is found almost exclusively in Canada. The remaining caribou subspecies, *R. t. pearyi,* lives on Canada's Queen Elizabeth Islands and surrounding islands and on Greenland's eastern seaboard, although the Greenland group is sometimes listed as *R. t. eogroenlandicus.*

Reindeer, like caribou, have also been broken down into subspecies, such as the Spitzbergen reindeer, *R. t. platyrhynchus;* the east Russian reindeer, *R. t. fennicus;* and the Scandinavian reindeer, *R. t. tarandus.* Adapting to the same range of environment as caribou, reindeer living under arctic conditions exhibit a lighter pelage than those living in forests. However, some reindeer will be piebald. Yet the feature most often cited as separating reindeer from caribou is size. Except for a few atypical populations, reindeer are slightly smaller than caribou—shoulder height averages a foot less. They also carry smaller racks of antlers than do caribou. In all, the differences between caribou and reindeer are not much greater than those differences separating the various caribou subspecies—keep in mind, reindeer and caribou are simply differing subspecies of *Rangifer tarandus.*

While their size may be the most noticeable feature separating the two, reindeer are better known for their domestication, a lifestyle not shared by caribou. Many reindeer still roam free, but thousands live as domesticated or semidomesticated members of managed herds. Such influence over a normally wild animal is possible only because reindeer—and, presumably, caribou—are very easily tamed. In a 1982 study of the Finnish Saami (Lapps), it was found that about 200 Saami families oversaw the migration of 100,000 reindeer over an area larger than the state of California. This command of reindeer is fairly recent; the actual taming of reindeer did not begin until about the fifth century, when individual reindeer were used to lure other reindeer to hunting parties. Since then, they have served as draft animals and as a source of various necessities, providing their keepers with leather, milk, meat, bone implements, and clothing. One of their major assets is their low maintenance. Reindeer do not need to be corralled or stabled during harsh weather, and they forage for themselves. For the Saami and the others living in the Arctic, reindeer have become an indispensable part of their lives.

Coyote • Dog • Wolf

Although we commonly regard the wolf, the coyote, and the dog[12] as three separate species, their ability to interbreed and produce fertile offspring somewhat undermines this distinction (the inability to successfully interbreed is usually considered a major indicator of distinct species).

Another unusual aspect of their interbreeding is that it is often volun-tary—that is, done without human prodding—which in some cases has produced remarkably large populations of hybrids. Of the three possible combinations of such pairings, that of wolves and coyotes has been the most widespread and successful.

In the latter part of the nineteenth century, coyotes *(Canis latrans)* from Minnesota began mating with gray wolves *(C. lupus)* in the north-ern part of the state and in western Ontario. These animals, with a bent toward hybridization, then moved eastward into territories once domi-nated by the wolf. As a result of its continued migrations, this hybrid (rec-ognized as the Eastern coyote, *C. latrans* var.) now inhabits much of New Brunswick and the New England states and is currently making inroads along Quebec's southern margin. Yet this mixing affected not only the coyote but also left its mark on the wolf. DNA analysis indicates that most of the wolves now roaming Minnesota and southwestern Ontario carry genetic traces of ancestral coyotes.

Due to lack of opportunity, mating between wolves and feral domes-tic dogs *(Canis familiaris)* is far less common. Most of the crossbreedings between the two have been the result of deliberate attempts to produce a strain exhibiting the more desirable traits of each. In the past, efforts by American Indians along these lines generally produced headstrong, if not dangerous, animals; however, they did make excellent sled dogs. Today, it is not uncommon for wolves to be bred with Alaskan malamutes or Siber-ian huskies simply for novelty. While producing a fairly mild-tempered animal, the resulting hybrid is still inclined to be shy of humans and dif-ficult to train and retains strong territorial and pack instincts. And although these undesirable characteristics have not deterred such crossbreedings, they have prompted some states to ban the possession of wolf-dog hybrids.

Coydogs, the offspring of (usually male) coyotes and (female) dogs, are seldom the result of deliberate crossings but rather of chance meetings between domestic dogs and the occasional coyote that frequents human settlements. At the outset these hybrids tend to be aggressive animals; however, given time to grow accustomed to humans, they usually become friendly, if not somewhat timid. Because of the lack of opportunity, wolf-coyote hybrids and coydogs rarely mate with each other. Furthermore, unlike the fertile offspring produced by a wolf and a wolf-dog cross, when a coyote and a coydog mate, the result is sterile pups.

Whether or not these various hybrids effectively blur the distinction between the three species, most authorities who have assessed the char-acteristics of purebred wolves, coyotes, and dogs still endorse the validity of separate species ranking. This is not surprising, considering that it has been 2 million years since the coyote broke away from the evolving wolf line, giving the two ample time in which to establish themselves as distinct species. Not so obvious is the domestic dog's ranking as *C. familiaris:* more than one authority has suggested it be considered a wolf subspecies or race.[13]

This animal is purposely identified as the domestic dog to distinguish it from the "dogs" of other genera. In fact, it is only through domestication that the domestic dog has attained status as a separate species. This "man's best friend" is widely believed to have been selectively bred from wolf strains some 10,000 to 14,000 years ago (possibly from the Asiatic wolf or a dingolike prototype), although there are indications that it may in fact be much older. DNA evidence that came to light in 1997 indicates that dogs may well have evolved from wolves 100,000 years ago. But even if it does bear a much older ancestry, today's domestic dog still retains many wolf characteristics—despite the many breeds that only marginally resemble wolves. Foremost among these shared traits is their nearly identical genetic profiles. Dogs and wolves also exhibit similar forms of behavior, social organization, vocalization, temperament, and body posture. This strong connection between the two is most evident in the wolflike physique of dogs such as the malamute, which has been left relatively untouched by experimenting breeders (most dog breeds, now numbering more than 400, were developed in the last 150 years through intensive selective breeding).

Whereas *C. familiaris* is the result of artificial breeding, other so-named dog species in the Canidae family arose naturally. There is the raccoon dog *(Nyctereutes procyonoides)* of Asia and Eastern Europe; the African or Cape hunting dog *(Lycaon pictus)* of Africa; the bush dog *(Speothos venaticus)* of South America; and the red dog, or Dhole *(Cuon alpinus),* that lives in India, Southeast Asia, and eastern Russia. Although these animals do resemble domestic dogs to varying degrees, they all possess characteristics that preclude their inclusion in *Canis.*

The short-legged raccoon dog, which actually resembles a raccoon as much as it does a dog, is an anomaly, with no other close relative in the family. It is distinguished by its underdeveloped carnassial (shearing) teeth and enlarged molars. The Cape hunting dog stands apart from those of *Canis* because of its articulating collarbone, lack of a tail gland, and missing dewclaws on its forefeet. Both the bush dog and the red dog are excluded because they have less than the 42 teeth normally found in canids: the bush dog is missing a molar on each side of its upper and lower jaws, and the red dog lacks a final lower molar on each side.

These differences among the canids are thought to have arisen in the distant past, the earliest occurring 7 million years ago, when the raccoon dog broke away from the others. Following this split, the wolflike canids and the major fox lines eventually came into their own (the gray fox had separated from the others much earlier). As the "wolf" line continued to evolve, other doglike animals began appearing. Among these, the coyote, the black-backed jackal, and the bush dog all made their debut 2 million years ago. At the same time, of course, the wolf was evolving toward its present state, forming at least three different species. One, *C. dirus,* the dire wolf, has since become extinct. Of the other two, only *C. lupus,* the gray, tundra, or timber wolf, continues to survive in large numbers. The remaining species, *C. rufus,* the red wolf, is unfortunately an extremely endangered animal that numbers less than 500. Most red wolves live in captivity, in rebreeding programs carried out by zoos and research facilities. The rest live as small, newly released populations in the wilds of several southeastern states. Although the red wolf is listed as a separate species, some scientists feel that it may be no more than a race of gray wolves, whereas others have suggested it may be a wolf-coyote hybrid.

There is one other canid called a wolf that does not fit into *Canis* because of its ancestry and unique features. This is the maned wolf *(Chrysocyon brachyurus)* of South America, a tall, very long-legged animal with large ears. Although it is more closely related to the fox than to the wolf, its true origin remains obscure. The aardwolf is another animal sometimes thought to be a wolf because of its name. This strange termite-feeding carnivore of Africa is not a canid but a close relative of the hyena.

Other than the 200-pound St. Bernard and the 165-pound mastiff, wolves are the largest animals of *Canis.* They generally weigh between 40 and 170 pounds and stand 26 to 38 inches high at the shoulder. Coyotes usually weigh between 12 and 40 pounds and have a shoulder height of only 18 to 26 inches. The domestic dog, having undergone innumerable breedings, is, of course, an incredibly varied species that ranges from the 2-pound miniature Chihuahua to the burly St. Bernard. And dogs vary not only in size but also in general appearance, a characteristic not inherent in wolves and coyotes—both are monotypic, showing little variation within their species. Examples of this wide variation in dogs abound. The Mexican hairless, as its name implies, is virtually barren of hair, whereas the Pomeranian has a long, thick coat; the ears of the bloodhound are large and pendulous, but those of the bull mastiff are short and upright; the legs of the dachshund are short in relation to its body length, whereas those of the Arabian greyhound are conspicuously long; the snout of the bulldog is severely flattened (falling far short of its lower jaw), but that of the Russian wolfhound is long and pointed.

These variations aside, as a species all dogs share characteristics that distinguish them from the wolf and the coyote. Probably most noticeable is their teeth. Whereas wolves and coyotes survive by seizing prey with their large canines and then slicing it up with well-developed carnassials, dogs, living as domesticated animals with no need to hunt, have developed smaller teeth, which in some breeds is accompanied by a loss of premolars.

This change to a more passive manner of feeding also resulted in a reduction in the cheekbones of dogs, leaving them less robust than those of wolves. No longer needed to anchor the substantial jaw muscles required by wild canids, the smaller cheekbones of the dog have a less prominent arch, which, in turn, makes for a larger orbital opening (the opening in the skull housing the eyes). This slight difference in orbital size is useful in distinguishing the skulls of dogs from those of other canids. Anatomists measure the angle formed between a horizontal line laid across the face of the skull and a line running from the upper edge of the eye socket through the socket's lower cheek edge. In dogs, this angle is between 53° and 60°; in wolves, it ranges from 40° to 45°.

Continued breeding and domestication are also thought to be responsible for the reduction and the shape of the auditory tympanic bulla in dogs. This bony projection at the base of the skull, thought to play a part in hearing acuity, is larger and rounder in the wolf and the coyote. It is a sensory tool vital to survival in the wild and best developed in the coyote. Also, the relative brain size has diminished slightly in the dog.

Although dogs originated from wolf stock, oddly enough, they developed dewclaws on their hind feet—a vestigial feature that wolves and coyotes lost millions of years ago. The continual breeding of dogs also left them with a prominent toe that varies with breed. In contrast, the largest toes on a wolf are always the middle two, and on coyotes it is always the outer toe. The stride pattern of dogs also changed as the animal was bred away from its ancestral wolf form. With the wolf and the coyote, the left back foot falls in the same line as the left fore foot, and vice versa. The dog, however, generally places its left back foot alongside the print of its left fore foot, and vice versa, leaving a trail of staggered steps. The foot placement of the wolf is attributed to the animal's deep and narrow chest, which holds its forelegs close to the body and in line with its rear legs.

The carriage of the tail is also a useful means of distinguishing dogs from wolves and coyotes. A wild canid has a straight tail that normally hangs down, but under stress or excitement it becomes rigid and is held at an angle of between 75° and 85° from horizontal. A dog under the same situation tends to curl its tail over its back, which is also a normal relaxed position in some breeds. This "sickle tail" position—a telling sign of the wolflike husky—is a juvenile characteristic in wolves and coyotes, which is lost as they mature.

However, a more significant distinction is in their different reproductive cycles. Whereas wolves and coyotes go into heat only once a year, dogs have two and sometimes three heats a year (the basenji, which goes into heat only once a year, is an exception). Living in sync with the female, male wolves and coyotes produce sperm only seasonally, when the female is in heat. In contrast, male dogs always carry sperm. The comparatively less-prolific nature of wolves and coyotes is in response to the needs of their newborn, which require about 2 years to reach maturity. Dogs, whose pups generally mature in only 6 to 9 months, have the good fortune of a more sexual and prolific life.

These characteristics, running true throughout each species, tend to give the impression that dogs and wolves are far removed from one another. That they are not is evidenced by the gradation in some of their shared characteristics. An example is the presence of the supercaudal gland in canines. In wolves and coyotes, this gland, lying on the topside of the tail close to the rump, continually exudes pheromones (whose exact purpose remains a mystery). In those dogs that have been bred away from their wolfish ancestry, this gland, while still present, no longer functions, whereas it is still quite active on the more wolflike dog breeds such as the malamute and the husky.

Although the canine family of today is represented by species spread throughout the world, all arose from ancestors that appear to have originated in North America during the Oligocene epoch. While foreign lands have served the canids well, one species has remained true to its homeland: the coyote. Strictly a New World animal, this very successful carnivore ranges from northern Alaska to Costa Rica, where it goes by several different names. Among the Aztecs it was known as *coyotl,* a name the Spanish adopted and later fashioned into its present form.

Compared to the wolf, the coyote is a far less robust canine. In addition to having a slender body that is only one-third to one-half the size of a wolf, the coyote's feet are smaller in proportion to its build (only a quarter the size of the wolf's foot), its snout is narrower and has a smaller nose pad, and its teeth are not as massive. However, the coyote's more pointed ears are larger than those of the wolf, and its distinctively black-tipped tail is far bushier, if not a bit longer. The manner in which it carries its tail when running is also telling. A wolf lets its tail float behind as it runs, whereas a running coyote always holds its tail down, sometimes tucked between the legs.

As a smaller animal, the coyote rarely attacks prey larger than deer. Typically, coyotes stick to a diet of small to medium-size mammals (rodents, rabbits, weasels, and livestock), vegetables and fruit, plus an occasional bird, fish, snake, or lizard. And they do not hesitate to scavenge the food of others, including that left unattended by humans. Any large game they do kill is usually a wounded or dying animal or one they can take down with the help of other coyotes. But such pack behavior is not typical of coyotes—most travel alone. If they should form a pack, it is usually temporary and small, with between 3 and 8 members. Wolf packs are usually more permanent and larger. The average wolf pack consists of 8 individuals, although some have been spotted with as many as 20 members.

Dolphin • Porpoise • Whale

"Use dolphin and porpoise interchangeably . . . and no one will bother to correct you"—an observation made in *National Geographic* over twenty years ago that still holds true today.[14] This casual attitude toward the proper use of the two terms, often creating unnecessary confusion, is unfortunate because dolphins and porpoises comprise two distinct families, whose differences, while not obvious, are fundamental. Several sources, including science, appear to be responsible for this uncertainty; so while these faulty notions may be unwarranted, they are understandable.

Many fishermen familiar with the dolphin *fish* (true dolphins, porpoises, and whales are mammals) customarily call dolphins "porpoises" to differentiate the mammal from the fish. Also, seamen have often misidentified dolphins as porpoises because of the dolphin's fondness for "porpoising": their propensity to leap out of the water as they swim alongside ships. Unfortunately, the designation "dolphin" is just as misused. Whether it is because dolphin species outnumber porpoise species 36 to 6, or because the most well known of these mammals is the bottlenosed dolphin, the public frequently calls porpoises "dolphins." But just as likely a source of confusion are the many popular books and articles describing these mammals.

Popular literature on dolphins and porpoises is replete with confusing references to the two. It is not uncommon to see both terms used for the same mammal or the features of one improperly ascribed to the other. To be fair, however, such misinformation likely comes from relying on various pieces of scientific literature, in which the two terms were given different definitions.

In examining the traditional terminology of dolphins and porpoises, K. S. Norris, one of North America's foremost cetologists (those who study whales and related mammals), redefined "dolphin" and "porpoise." "Dolphin," he contended in 1961, should be applied only to a single species, the common dolphin. Some "whales" in the family retained their names, while all other family members he labeled porpoises. Ultimately, this renaming did not prevail, nor did various other labelings proposed by fellow scientists. But as a consequence of the unsettled definition of the two terms, many scientific studies handled "dolphin" and "porpoise" very cautiously, often using them in only a generic sense and relying on scientific names for more precise identification.

Unfortunately, it appears that authors of popular literature picked up on these writings and, using common names rather than their scientific labels, not only confused dolphin and porpoise features but combined outdated notions with current views. This gave the erroneous impression that the scientific community didn't seem to know what it was talking about. Consider the following statement from a prominent scientific encyclopedia: "Porpoise means 'bottle-nose' as contrasted with the beak-type nose of the dolphin."[15] First of all, "porpoise" does not mean "bottle-nosed" but "pig fish," a name chosen because of the mammal's resemblance to swine (stout form, naked skin, and noisy exhalation). Second, "bottle-nosed" is never applied to porpoises, only to certain dolphins. Another statement reads in part: "Dolphins are small toothed whales. . . . There are many species. . . . One of the more peculiar is the narwhal. . . ."[16] The narwhal, with its uniquely twisted shaft jutting out just above its upper lip, is not a dolphin but a true whale, classified with the beluga (white whale) in their own family.

Some authorities group cetacean families into superfamilies. In these schemes, the families of dolphins, porpoises, beluga whales, and narwhals make up the superfamily Delphinoidea. Within this framework, and in the broadest sense, the narwhal could be called a dolphin. Yet in the work cited previously, no superfamily level was established, so the reference to the narwhal as a dolphin is erroneous.

Fortunately, in the last few years popular writings that address the differences between dolphins and porpoises have become more dependable, as has the application of their respective names—due, no doubt, to the use of more reliable information. Today, cetologists, although not entirely in agreement, do seem to be using the common names "dolphin," "porpoise," and "whale" more uniformly. Yet so many different classification schemes (from the rank of family or superfamily down) remain that the exact relationship between these mammals is unclear. Therefore, while the following ordering of cetaceans is based on current knowledge, it is not the only one.

Whales, dolphins, and porpoises comprise the order Cetacea, which is divided into two suborders: Odontocetes, the toothed whales, and Mysticetes, those with baleen (horny plates on each side of the upper jaw). Because "whale" is the English equivalent of the Latin *cetus*, from which the order derives its name, when one speaks in general terms, "whale" is frequently used for all members of the order. But to be precise, the term should be reserved only for those cetaceans that are neither dolphins nor porpoises. Unfortunately, a few dolphins do have the word "whale" in their name, such as the killer whale, the longfin pilot whale, and the melon-headed whale. These are all true dolphins of the dolphin family Delphinidae. Aside from the fact that two of the three are larger than average, the term "whale" in their names is trivial.

Except for the dolphin family (Delphinidae), the porpoise family (Phocoenidae), and the questionable river dolphin family (Platanistidae), all other toothed cetaceans are considered "true" whales—as are all baleen whales. It is usually quite easy to identify whales because of their size, but for the mammalogist, more than size alone is needed to distinguish them from the dolphins and the porpoises.

If we disregard the river dolphins for the moment and, for present purposes, designate the Delphinidae dolphins and the porpoises as D&Ps (in some classification schemes, porpoises are placed in Delphinidae), three families of toothed cetaceans remain: beaked whales, sperm whales, and Monodontidae whales (the narwhals and belugas). Two features separate the beaked and sperm whales from the D&Ps and Monodontidaes. One is a difference in the articulation of the posterior ribs, and the other, their throats. Only the beaked and the sperm whales have throats with deep external grooves. Separating the D&P families from Monodontidae whales requires an examination of their first and second cervical vertebrae: only those of the D&Ps are fused together. The dorsal fin may also be a distinguishing feature. Monodontidaes have none, whereas, with the exception of the finless porpoise and the right whale dolphins, all D&Ps carry such fins.

Although the taxonomy of the dolphin family Delphinidae is far from settled, that of the river dolphins, Platanistidae, is fairly well established (some authorities divide the river dolphins into three families under the superfamily Platanistoidea). Platanistidae has five species, which inhabit the freshwater rivers of South America, China, and several countries bordering the Bay of Bengal and the Arabian Sea. These are such distinctive cetaceans, with characteristics so unlike those of the Delphinidae dolphins, that the relationship implied by their common name can be misleading. Although smaller than Delphinidae dolphins, the river dolphin has an exceptionally long and slender beak and a prominent forehead, and none of its cervical vertebrae are fused. It also carries 8 pairs of double-headed ribs, compared to the 4 or 5 pairs borne by the dolphin. Two species carry token dorsal fins, and two have only rudimentary, lensless eyes. Truly unusual cetaceans and seldom considered to be as closely related to dolphins as are the porpoises, we call them dolphins only because of their superficial body resemblance to the Delphinidaes.

In popular works where a distinction is made between dolphins and porpoises, it is often stated that dolphins have beaks, whereas porpoises do not. It is true that porpoises have no beaks, but neither do the Risso's dolphin and the killer whale. Neither is it true that all porpoises are smaller than all dolphins, just as not all dolphins are smaller than all true whales. There is, however, a fairly consistent difference in their shapes: dolphins tend to be more slender and streamlined than are porpoises. But where their fundamental differences really lie is in their teeth and in the sounds they make.

Dolphin teeth, like those of whales, are either peg shaped or, more commonly, pointed (as are those of the river dolphins). Porpoise teeth are spadelike, flattened laterally, and set at an angle in the jaw to help in shearing food. And it is this single feature, the difference in their tooth form, which was the principal consideration in establishing the two families.

The sound these mammals make has two purposes: echolocation and communication. It is not a feature that taxonomists use for classification, but its particular form in dolphins differs from that in porpoises. Porpoises, as do sperm whales and river dolphins, seem to produce only clicks; whereas dolphins, besides making clicks, also produce whistle calls and pulsed squeals. Other characteristics, such as the shape of the head and the contour of the dorsal fin and the flipper, may be somewhat distinguishing, but because they do not occur in all species, they cannot be considered defining.

Dugong • Manatee • Seacow

Sirenia, the order of the dugong, manatees, and seacows, consists of only five species (four confirmed living). Other than Cetacea, the whales, it is the only other order of wholly aquatic mammals. While a number of differences separate the two orders, the most noteworthy is that sirenians are herbivores and cetaceans are carnivores.

Although "seacow" is the name of a particular sirenian, the term is often encountered as a collective or an individual label for the dugong and the manatee as well. While these uses of "seacow" are questionable, some zoologists feel that they do have merit for a couple of reasons. First of all, the seacow (Steller's seacow, *Hydrodamalis gigas*) has been extinct since 1768—or so it appears—so any living sirenian called a seacow would not be confused with Steller's seacow. Second, applied to any of the sirenians, the image evoked by the name "seacow" is quite apt. All are slow-moving, rotund mammals that remind one of a large floating cow. This bovine character is well recognized by other cultures. The Spanish call the manatee *vaca marina* (sea cow), and the Portuguese term means "fish cow." Because some authorities still have reservations as to the seacow's extinction—a possible sighting was made in 1962—it is included here along with the dugong and the manatee.

Seacows, like all sirenians, are thought to have been distant relatives of the elephant. Nowhere near as large as the typical whale, they were still enormous animals. Various estimates of head-to-tail length range from 18 to 30 feet, and their weight is judged to have been between 7,000 and 13,000 pounds—surpassing that of the average elephant. Usually regarded as a member of the dugong family, the seacow possessed several notable characteristics. It weighed three to six times more than any other sirenian; its only food was a seaweed form of algae; lacking teeth, it processed this food with horny plates covering its gums and palate; it lacked finger bones; and its skin was very wrinkled and toughened, almost like tree bark.

Seacows, keeping to colder waters, developed an extremely thick layer of insulating fat that also served as a food reserve. All this blubber attached to a good-sized frame, plus a very docile nature and slow breeding, eventually led to the seacow's apparent demise. Discovered in the Bering Sea in 1741, it was so aggressively hunted by whalers and seal hunters that within twenty-seven years, it appeared to be extinct—probably one of the shortest encounters any animal has ever had with man. The dugong (from the Malayan *duyong*) and the manatee (from the Taino name *manati*—the same Caribbean people who gave us the word "maize") evaded a similar fate only because their smaller size made them less profitable.

Compared to the three somewhat bloated-looking species of manatees, the lone dugong species, *Dugong dugon,* is almost streamlined. Unlike the cold-loving seacow, the dugong's preferred seas include the temperate waters off east Africa, the Red Sea, the Indian Ocean, and the waters above Australia and eastward into the Pacific. Never venturing very far out to sea, it inhabits shallow coastal regions and will frequently explore estuaries that have not become overly diluted with fresh river water.

The dugong's principal food is sea grass, which it chews with horny pads that cover its palate. Unlike the seacow, it does have a few teeth in the rear of its mouth. The male dugong also carries a pair of short upper incisors called tusks that jut forward and are used in mating, but none of its teeth play a significant role in feeding. Also, in contrast to the seacow, whose mouth faced forward, enabling it to easily eat tall seaweed, the dugong's mouth angles downward for bottom grazing. To digest their food, the dugong and the manatee do not have sectioned stomachs or rumens, as do cows, but employ very long (over 100 feet) intestines, designed to hold great quantities of food.

Whereas the dugong swims the Indo-Pacific seas, manatees (*Trichechus* spp.) occupy the waters along western Africa and the Americas from Virginia down to Salvador, Brazil. Somewhat more far ranging than the dugong, they have been spotted several miles out to sea and hundreds of miles up fresh river waters—particularly the Amazon River and those of western Africa.

The manatee survives in these varied habitats because it feeds not only on marine bottom grasses but on seaweed and assorted freshwater plants and algae as well. Usually an underwater diner, it will occasionally surface to eat floating or bank vegetation. To assist in foraging, it employs its flippers to grasp vegetation—their only use. In contrast, the dugong uses its relatively smaller flippers for swimming and only rarely for feeding. The tuskless manatee, not equipped with a rough palate to chew its food, uses a unique set of molars in the rear of its jaw for eating. Aligned in rows, as the molars wear down, they move forward with age and are continually replaced by new teeth from the rear.

In outward appearances, the dugong and the manatee are very similar and could easily be mistaken for each other but for one notable feature. The tail of the dugong (and of the Steller's seacow) resembles those of whales; it is notched in the center and fans outward, whereas the manatee tail resembles an unnotched oval spatula. These anatomical variations are but a few that separate the dugong and the manatee. Others range from variations in skull construction to minor differences in external features. The main reason for the diversity between such seemingly similar mammals lies in the time they have had to evolve. It is estimated that around 50 million years ago (during the Eocene era), the evolutionary lines of the dugongs and the manatees went their separate ways. Over this period, certain features of the sirenians changed in response to disparate needs, but all were retained in a very successfully adaptive body form.

Among the differences, one in particular is very peculiar. In the entire class of mammals, only the manatee and the two-toed sloth have but six neck vertebrae—all other mammals have seven. While extraordinary, this is not accorded much evolutionary significance and is seen as more of a chance anomaly. Other distinguishing features are as follows:

	Dugong	Manatee
Weight	500–2,000 lbs.	770–3,550 lbs.
Length	3–13 feet	7.5–15 feet
Color	Slate or bluish gray	Light to dark gray or brown with pink patches
Faces	Smaller, with many bristles on the snout	Larger, with fewer bristles
Nostrils	Do not close completely	Close tightly
Flippers	No nails	Three, except for the Amazonian species with none
Body hair	Virtually none, except on face	Sparsely scattered
Muzzle	Faces downward	Faces forward

Elk ● Moose ● Red Deer ● Wapiti

Before the 1980s, when the North American elk was listed as a specific species *(Cervus canadensis),* the term "elk" was part of a confusing mix of labels involving three different animals. In Europe, *C. canadensis* was not called an elk but rather a "wapiti," a Shawnee Indian name meaning "white deer." Europeans were compelled to use another name because they had already bestowed "elk" on a quite different animal, *Alces alces,* a species better known in North America as the moose. Adding to the confusion, two Asian subspecies of red deer *(Cervus elaphus)* are sometimes dubbed "wapiti," leaving the three species with the following combination of now, fortunately, obsolete designations.

Scientific Name	North American Name	European Name
Cervus elaphus	Red deer	Red deer (two as wapiti)
Cervus canadensis	Elk	Wapiti
Alces alces	Moose	Elk

Why the colonists of North America used the name "elk" for an animal that is similar to the red deer of Europe when it already denoted the European moose is unclear. The moose, a large (1,300 lb., 7-ft. at shoulder height) animal with broad, palmate antlers and a distinctive snout, is quite unlike the North American elk, a smaller (800 lb., 5 ft.) animal that carries branched antlers and has a more conventional muzzle.

The elk and the European red deer, two very closely related ungulates, share their genus *Cervus* with several other deer scattered across the northern hemisphere. These related deer, living in very remote regions, include the hangul of Kashmir, the shou of Tibet and Bhutan, and the Barbary stag of North Africa. Not only do these particular animals share the same genus, but they also belong to the same species: *C. elaphus*—as does the North American elk. The elk of North America (hunted to extinction in the central and eastern regions of the continent during this century) is no longer considered a separate species *(C. canadensis)* by most authorities and has been reassigned to *C. elaphus* as *C. e. canadensis*. Its very close living relatives, the Olympic or Roosevelt elk, the Manitoba elk, the Rocky Mountain elk, and the Tule elk, were also placed in *C. elaphus* as separate subspecies. This inclusion in *C. elaphus,* however, has not changed their common name; North Americans still refer to them as elk, and Europeans continue to call them wapiti.

In 1987 zoologists took another step in classification when they combined the 25 subspecies of *C. elaphus* into three distinct subgroups: those closely resembling the European red deer, those resembling the North American elk, and those living in the lower regions of China and central Asia. The ancestors of this last group of more primitive subspecies are thought to have sired the red deer and the elk subgroups. But this elk subgroup, in its emigration to North America, left many of its members behind—close relatives that still inhabit Korea, Mongolia, Manchuria, and the Tien Shan Mountains. These particular deer continue to flourish in Asia and to round out the "North American elk" grouping of thirteen subspecies.

As subspecies, red deer and elk do closely resemble each other, yet if one focuses on their subtle differences, it is not difficult to tell the two apart. The red deer of Europe, North Africa, and southwestern Asia tend to be smaller (4½ ft. high at the shoulder) than elk are, and bear smaller and slightly different-configured racks of antlers (the largest span is about 43 inches, versus 66 inches on elk). They also have proportionally smaller white patches on their rumps, and, unlike the darker colored elk, they retain their reddish color during the summer months. The mating call of each is very different, the red deer producing a deep roar and the elk letting out a high-pitched bugle. Both mate in early autumn but have slightly different gestation periods. Gestation in the European red deer lasts about 235 days, whereas an elk may go an additional month before giving birth—both usually bearing a single offspring.

At one time elk ranged over much of North America—possibly all 49 of the continental states—spanning the country from coast to coast and reaching from the Yukon down to Mexico. Their population, estimated to have once been around 10 million, slowly fell over the last two centuries to the guns of hunters and settlers. Today, elk number no more than 700,000 and can be found in only 12 states and 3 Canadian provinces. Most live in the Rocky Mountains, the Pacific Northwest, the Sierra Nevada mountains, and south and central Canada.

Red deer continue to live across Europe and Asia in more-or-less isolated subspecies groups. Ranging from the British Isles and Scandinavia to North Africa and into eastern Tibet, many populations are quite small. One of the smallest, *C. e. corsicanus,* is struggling for survival on the island of Sardinia (it is currently being reintroduced into Corsica). Because the range of red deer is more extensive than that of elk, their size is more varied. *C. e. hippelaphus* of Western Europe weigh in at 660 pounds, whereas *C. e. corsicanus* weigh no more than 167 pounds.

The moose, distinctive as it is, should never be confused with any other animal—red deer and elk included. Its most distinctive features have already been mentioned, but a few others are noteworthy. The largest of all deer (elk come in second), a moose weighs between 880 and 1,750 pounds (females weigh 25 percent less than do males) and carries an enormous rack of antlers (the record span is just short of 7 feet). Its drooping, fleshy muzzle allows it to easily strip leaves from branches, as well as to root vegetation out of marshy ground. With a heavy body crowned by a prominent shoulder hump, its long legs take on a particularly spindly appearance, altogether leaving the moose looking somewhat clumsy—which it is not. It also carries a unique "bell" of skin and hair dangling from its throat that may reach 20 inches in length. The coat of the moose, which is lighter in winter than in summer, is dark brown, with a lighter brown on its underside and legs (in the winter, elk will often develop a light-colored body with dark legs).

In September and October, the male moose calls for a mate with a loud bellow. If he is successful, the mated female goes into gestation for about 260 days, which typically results in a single birth. Developing quickly, both sexes can breed when 1.5 years old. Those living in Europe do not grow as large as North American moose do and have correspondingly smaller racks.

Ermine ● Ferret ● Fisher ● Marten ● Mink
● Polecat ● Sable ● Stoat ● Weasel

"Weasel" is not simply the name of certain species; it is also the informal label for an entire family of carnivores (Mustelidae) and one of its subfamilies (Mustelinae)—Latin words that are based on the skunky odor these animals emit. These informal uses of "weasel," which can be highly misleading, obscure the fact that true weasels belong to only one of Mustelidae's five subfamilies—the other four being better known for their skunks, badgers, otters, and ratels. The same is true of the "weasel" subfamily Mustelinae; it contains not only weasels but martens, fishers, minks, polecats, sables, zorillas, and wolverines as well. The broader, more encompassing use of "weasel" as a family and a subfamily label is simply a convenience—chosen because weasels were at one time the most widely known mustelids. More cautious authorities call these categories "weasels and allies," a label still deferring to the weasel's wide renown.

Actually, weasels cut across genus lines. Whereas most belong to the genus *Mustela*, they also occur in at least three other genera. Yet when one considers that almost half the species of *Mustela* are not called weasels, it becomes apparent that "weasel" is not a particularly defining term. Accompanying the weasels in *Mustela* are ermines, minks, ferrets, polecats, and the kolinsky. The only feature uniting the various animals known as "weasels" (as well as the rest of those in *Mustela*) is their body shape. Except for the huskier North African weasel of *Poecilictis*, each one has a "weaselly" appearance: a long neck and a slender body supported by short legs. Among the family's remaining members, the heftier marten, the zorilla (an animal often mistaken for a skunk), and the much larger wolverine all lack the weasel's typical sleek appearance. The following genera of the subfamily Mustelinae illustrate the relationship of the weasels to their allies.

Genus: *Mustela*
    8 species called weasel
    2 species called mink
    2 species called polecat
    2 species called ferret
    1 species called ermine
    1 species called kolinsky
    1 species unnamed (new)

Genus: *Poecilictis*
    1 species called weasel

Genus: *Poecilogale*
    1 species called weasel

Genus: *Lyncodon*
    1 species called weasel

Genus: *Vormela*
    1 species called polecat

Genus: *Martes*
    6 species called marten
    1 species called fisher
    1 species called sable

Genus: *Galictis*
    2 species called grison

Genus: *Ictonyx*
    1 species called zorilla

Genus: *Gulo*
    1 species called
    wolverine

Because weasels occupy four genera, it is difficult to cite specific "weasel" characteristics; it is more productive to consider weasels in terms of what they are not by focusing on the distinguishing features of the other mustelids. A few notes on weasels are appropriate, however.

In general, weasels tend to be the smallest members of the family, ranging from the tiny 6½-inch-long (head and body length) least weasel *(Mustela nivalis rixosa)*, to the 14-inch-long Patagonian weasel *(Lyncodon patagonicus)*. Of the 12 species traditionally regarded as weasels, only one lives in Europe, the European common weasel *(Mustela nivalis):* a small creature better known as, simply, "weasel." Ranging throughout Europe, as well as northern Africa and across Asia above the Himalayas, it varies considerably in size and coloring. The least weasel, often considered a subspecies of the European weasel, is one of two species found in North America. The other, the long-tailed weasel *(Mustela frenata),* is the larger of the two and lives as far south as Bolivia.

The two mink species, one native to North America and the other a European resident, differ most notably in size. It is believed that the smaller European species distanced itself from its North American ancestor some 10,000 years ago, when the Bering land bridge disappeared. As the only semiaquatic members in its family, minks have slightly webbed feet that aid in hunting underwater. To insulate the body against the water's cold, their fur has long guard hairs that overlie a dense layer of insulating shorter underhairs. Secreted oil coats these hairs to keep them dry and imparts a luster to the fur, for which minks are noted and valued. Except for a white chin and a spotted throat, the mink's soft fur is a uniform, rich, blackish-brown color (weasels living in the same part of the world can be distinguished by their light underparts). And unlike the coat of the ermine and most weasels, that of the mink does not change color with the seasons. Its long, black-tipped, bushy tail accounts for about a third of its total length.

Although "polecat" is used in some parts of the United States as a disparaging comparison to the onerous skunk, skunks are not polecats—

a misunderstanding that has caused long-standing confusion. In fact, polecats do not live in North America. Not nearly as smelly as a skunk, the polecat is named for an equally offensive practice the French called *pouchet,* or "chicken killer." While polecats are included in *Mustela,* as shown, some authorities feel that they deserve their own genus and place them in *Putorius.*

Polecats are not wide-ranging mustelids, preferring instead to keep to the northern forests of Europe and the steppes and semideserts of Russia and Mongolia. The European polecat is about the size of a mink (about three pounds), and the steppe species is about a third heavier. The marbled polecat of *Vormela* is much smaller, weighing in at a pound and a half. This is twice the size of the average weasel, a more agile creature. Like other members of *Mustela,* the marbled polecat has a dark mask across the eyes, but its black coat is uniquely marked with stripes. The fur of the European species is more variable, ranging from black to buff, but always with dark belly fur. The steppe polecat also has darker fur on its belly, but the rest of the coat is reddish-brown.

Although the European polecat is considered by some to be a species in its own right, many zoologists feel that it should now be listed as a subspecies: *Mustela putorius putorius.* This is based on the belief that the European polecat gave rise to the ferret of Europe, an animal long included in *Mustela* and often designated as *M. p. furo.* Authorities who dispute this reclassification contend that it is more likely that the European ferret had its origins in North Africa and is unconnected to the polecat, which they classify as *Putorius furo.* Still others speculate that ferrets were bred from the steppe polecat. But whatever its origins, most scientists feel that the ferret of Europe is an offshoot of some form of albino polecat, an animal with characteristic red albino eyes and a light, buff-colored coat. The ferret, about the size of the European polecat, has two litters a year, whereas the polecat has only one. Like the name "polecat," "ferret" has a similar origin, coming from the Old French *fuiret,* for "thief."

The black-footed ferret, always listed as a separate species, *M. nigripes,* is an exclusively North American mustelid—one of the rarest animals in the world. Named for its unique black lower legs and feet, it is also distinguished by a black face mask and a black-tipped tail, all of which are set against a buff-colored coat with black-tipped hairs. Thought to have evolved from the steppe polecat after migrating to North America during the Pleistocene era, it is the only representative of the polecat line in the New World. Unlike its forest-dwelling relatives, the smaller 1½-pound black-footed ferret prefers the open country of the prairie dog.

The ermine, *M. erminea,* is a much smaller mustelid, weighing no more than twelve ounces. Because of the ermine's size, it takes many ermine pelts to make a garment. This, coupled with its beautiful white coat, makes it the most valuable of all furs. Its name is the shortened form of the Latin *armenius,* for "mouse." Not as clear is the origin of its alternate name: stoat. Depending on regional preferences, either name may serve as its principal designation.

As individual terms, each reflects the seasonal difference in color of the animal's coat: "ermine" is usually used during winter when it is white, and "stoat" during summer when brown. However, this is true only of those animals living in cold winters, where white fur against a background of snow serves to camouflage the ermine. In warmer climates, the ermine—then best called a stoat—remains brown year round. On a climatic borderline, the animal's coat may never become completely white or brown but may remain piebald throughout the year. This is particularly true of those in England and North Africa. Throughout the rest of its range, which includes Northern Europe, Eurasia, Japan, and North America, it displays more typical seasonal changes.

The ermines of North America are sometimes called short-tailed weasels, a designation meant to set them apart from the long-tailed weasel, which also changes color with the seasons. The two can be easily distinguished by the ermine's smaller size (3 ounces versus 8 ounces) and its shorter, black-tipped tail.

The remaining mustelids under consideration, the marten, the fisher, and the sable, all belong to the genus *Martes.* Taxonomically, their four extra premolars and the cusp pattern of their lower carnassial teeth distinguish them from species of *Mustela.* The marten, the fisher, and the sable are also much larger animals, only moderately elongated, and with skulls more like that of a dog. All have beautiful fur and long, bushy tails. Except for the 5- to 13-pound fisher, they generally weigh between $3\frac{1}{3}$ and $4\frac{1}{2}$ pounds. Typically arboreal, the stone marten and the American marten feel more at home among tree branches than on ground; conversely, the sable does not do well in trees, preferring to spend most of its time on the forest floor.

Although some marketed furs are called "American sable," the true sable *(M. zibellina)* lives only in northern Asia and the northern islands of Japan. Because its beautiful dark-brown to blue-black fur with tips of silver is so highly prized, furriers have sometimes tried to capitalize on its name by passing off the fur of the American marten *(M. americana)* as virtually identical with sable. This, in turn, led to the misbelief that sables were indigenous to North America. To distinguish the two pelts, most furriers call the fur of the true sable Russian sable or Siberian sable. The sable and the American marten, although close relatives, are readily distinguishable by the latter's reddish to yellowish-brown coat. The sable's tail is also much shorter, and the yellowish "bib" on its throat is not always distinct, unlike the marten's.

The American marten is also known as a pine marten, a name borrowed from the pine marten, *M. martes,* of central and Northern Europe. Not as closely related to the European marten as it is to the sable, the American marten is nonetheless outwardly very similar to the true pine marten. Whereas the American marten tends to have a more golden-brown coat and is slightly smaller, the coloring and size of the American and the pine marten fall within similar limits. Both also are very arboreal, hunting and playing among the conifer branches of northern forests.

The other martens in the genus are very similar. The stone marten of south and central Europe is distinguished by its large white bib that extends onto its forelegs. The Japanese marten of Japan and Korea and the Nilgiri marten of India are more yellow-brown to dark brown. The largest of all the martens is the Southeast Asian yellow-throated marten. Just as heavy as the fisher, it is a slightly shorter animal, with a more yellow-brown coat and a yellow to orange bib.

The fisher, *M. pennanti,* covered in medium- to dark-brown fur with a cream-colored bib, carries splashes of gold on its head and shoulders. About the size of a fox, this large North American mustelid hunts not only small prey but regularly goes after porcupines and occasionally young deer. Although it will feed on fish and is thought by some to be named for its penchant for stealing from fish traps, the fisher's name is more likely a corruption of the European terms *fitchet, fitche,* or *fitchew,* which meant "nasty" and was used to describe the polecat. Authorities speculate that American settlers fashioned "fisher" from these names, which were alternate names for the European polecat (a smaller animal but one with a similar build), for what they thought was either a larger North American version of polecat or, at the least, a close polecat lookalike.

The fisher, more closely related to martens than to polecats, is still easy to distinguish from martens. Two to three times larger than the American marten (the average size of a male fisher is 3½ feet long and it weighs 10 pounds), the fisher has correspondingly more robust dentition. More important, its teeth are unique; the upper carnassial teeth have an exposed rootlet—a diagnostic feature for old skulls. Another telling feature of the fisher is its four teats, half the number of the marten's. Probably the most arboreal of all the mustelids, the fisher, hidden in the upper reaches of a tree, can sometimes be recognized only by its screams and hisses. The scream of the marten is more a squeak than a hiss and is higher pitched.

Goat ● Sheep

More than a grain of truth resides in the old simile that trying to distinguish very similar objects from one another is often like "trying to separate the sheep from the goats." However, *most* sheep and goats can easily be distinguished from one another by noting several conspicuous features, the most obvious being the shape of their horns. It is only when one considers all the members of their subfamily, Caprinae, some of which exhibit a mix of sheep and goat features, that the distinction between the two quickly evaporates.

Caprinae is often called the goat-antelope subfamily, a reference to the ties its members have to others in its family, Bovidae. Evolving a basic body form somewhat similar to that of antelope but with shorter legs, caprid ancestors date back at least 30 million years. Yet it was not until the early ice ages that this once fairly homogeneous group is believed to have truly begun diversifying. Some, preferring the cold of the north, became longhaired, huge (770 lbs.), cattle-like musk oxen; others, like the goral of Southeast Asia, adapted to steep mountainous cliffs by becoming smaller (90 lbs.) and sporting only token horns. Most, however, evolved into medium-sized (200 lbs., 3½ foot tall) bovids that either became large-horned, wide-ranging grazers like the Himalayan tahr, or remained small-horned, territorial browsers such as the serow.

As they evolved, the caprids slowly split into two generally recognizable forms: those with "sheep" characteristics and those with "goat" characteristics. Stuck between these extremes—bearing features of both forms—the Barbary sheep and the blue sheep appear to be surviving anomalies formed at the juncture of this relatively recent split. Yet these two are not the most primitive of the caprids. That honor goes to the four small-horned species that make up the tribe Rupicaprini: the chamois, the goral, the serow, and the mountain goat. These are the so-called resource defenders, species generally leading solitary lives among the sparse shrubs and trees of steep slopes. The larger-horned musk ox and the takin make up the second tribe of Caprinae: Ovibonini. Both of these animals, believed to be offshoots of the rupicaprids, are wide-ranging grazers, the musk ox adapting to the sparse vegetation of the tundra, and the takin preferring alpine bamboo forests. The third, and last, tribe is Caprini, which is also considered an offshoot of the rupicaprids. Called the "sheep and goat" tribe, its members are highly gregarious grazers and browsers that live in locales ranging from low-lying grasslands bordering deserts to sparsely vegetated alpine meadows.

Most zoologists, despite their misgivings over the relationship between sheep and goats, do agree on the foregoing division of Caprinae, as well as Caprini's five genera. Caprini, consisting not only of those creatures that are unmistakably sheep and goats, also contains the Barbary and the blue sheep, the two questionable species. Taxonomically, the Caprini tribe breaks down into the following genera: *Ovis,* the sheep; *Capra,* the goats; *Hemitragus,* the tahrs; *Ammotragus,* with one species, the Barbary sheep; and *Pseudois,* also containing a single species, the blue sheep. Although zoologists do not regard the "sheep versus goat" issue as particularly crucial, they do strongly disagree on the systematics of *Ovis* and *Capra.* And, unfortunately, determining the correct relationships between the species of each genera and their classification has generated so much debate and revision that any taxonomy of *Ovis* and *Capra* remains open to question.[17] However, as distinct genera, *Ovis* and *Capra* remain clearly defined by well established characteristics. Some of these features can also be found in species belonging to the other three genera *(Hemitragus, Ammotragus,* and *Pseudois).*

As mentioned, probably the most notable difference between sheep and goats involves their horns, which are most pronounced in the males of both genera. In the sheep, these typically emerge close together from the top of the head, each corkscrewing rearward in a spiral that flares to the side. Their tips will end up pointing either forward or toward one another behind the head. The goat has much smaller, scimitar-shaped horns; spaced well apart on the head, they gently arc backward without twisting (the markhor, with large spiraling horns, is an exception). The goat also has a short and narrow skull with a convex forehead, whereas the sheep, although bearing a narrow nose, has a broader skull and a concave forehead. A more conspicuous feature of the male goat is its small chin beard, a feature absent on sheep. Goats also have heavily calloused knees, whereas the knees of sheep are covered with hair. If one looks at their rears, many sheep species, unlike goats, have large patches of light-colored hair on their rumps. The rump of the goat, however, does sport an upturned tail whose underside contains a scent gland. Such scent glands are lacking on the dangling tail of the sheep, which instead carries its scent glands on its hooves and its face. The strong stench characteristic of goats, particularly in the males, is often blamed on their scent glands, but much of this malodorous trait is really a result of their urinating on themselves.

Behaviorally, goats are more independent than sheep and appear to be more intelligent. A goat evading a predator will purposely take a route harboring the greatest obstacles to its pursuer. To reach edible leaves, some will climb into the branches of trees, and if this is not possible, they have been known to stand on the backs of obliging cattle to extend their reach. Unlike goats, sheep seldom browse but prefer grazing on grasses, sedges, and forbs. Sheep are more mellow and nonconfrontational and do not mount a dramatically vigorous defense when their lives or territorial rights are threatened. And, after being domesticated, they hesitate to return to the wild if given the opportunity, whereas goats will quickly become feral, easily adopting an independent lifestyle.

Many of these traits also appear in species of other tribes and genera, in particular the Barbary sheep and the blue sheep, both of which tend to be more goatlike in their behavior. The Barbary sheep, or aoudad, which has earned a reputation for going for long periods without water, roams the high mountain deserts of North Africa. The blue sheep, or bharal, a resident of the Himalayas and China, prefers steep slopes that overlook high cliffs.

Although the Barbary sheep is termed a "sheep," and analysis of its blood serum protein shows it to be most closely related to sheep, outwardly it resembles a goat as much as a sheep. Like a goat, it has a short tail with a naked underside containing a scent gland; it lacks face and foot glands; there is no rump patch; and it has characteristic long soft hairs on the throat, chest, and upper legs. Unlike a goat, the Barbary sheep has no beard, it does not smell, and its long horns sweep backward, ending in tips that face each other. A biochemical assessment made in 1977 found that the Barbary sheep has 5 characteristics exclusive to sheep, 5 exclusive to goats, 7 shared by both, and 10 that are unique.

The blue sheep, whose genus name *Pseudois* means "false sheep," is another curious mix of sheep and goat features. It has a typical goat tail and no facial or foot glands, and its horns, while large, turn sideways and then arch back without twisting, as in goats. Like sheep, it is beardless, lacks calluses on its knees, and does not emit a pungent odor. Unlike the Barbary sheep, the blue sheep appears to be more closely related to goats.

Great Cats • Lesser Cats

This ordering of the cat family into two groups is sometimes labeled the large cats and the small cats, or the big cats and the small cats. Regardless of the label, the distinction between the two groups lies mainly in the differences that describe the cat family's two major genera: *Felis,* the lesser cats; and *Panthera,* the great cats. The only questionable member is the cheetah. Although this unique cat, which occupies its own genus, *Acinonyx,* is considered by some to be most closely related to the cougar, a lesser cat, it is usually placed in the great cat group.

The origin of the cat family, Felidae, has been traced to the neofelids or "true cats," which existed 25 million years ago. Springing from this early progenitor, cats underwent various evolutionary adaptations and branchings that eventually culminated in today's 40-some species. But none of these branchings are well understood or universally agreed upon. Because fossil evidence linking the different forms of cats is scarce and often ambiguous, most of the cat family's presumed lineage is based on considerable supposition and remains rather tenuous. This shaky status extends even to today's two main groups of cats: the great cats and the lesser cats. And, unfortunately, there is little evidence showing what form their common ancestor may have taken or when the two groups eventually came into their own. Even among the cats of today, the relationships connecting various species are poorly understood. So, even though some cats bear features considered characteristic of the other group, these exceptions are not felt to be strong enough to establish a specific link between the great cats and the lesser cats.

The labels "great cats" and "lesser cats" are, of course, not value judgments but a reflection of the average size of the cats forming each group. The great cats, the lion, the tiger, the leopard, the jaguar, the snow leopard, and the cheetah, are all large animals. They range from the smallest, the cheetah, which can weigh up to 158 pounds, to the largest, the tiger, which can weigh in at 660 pounds. But it would be a mistake to think these cats are all larger than any of the lesser cats. The cougar, a lesser cat that weighs as much as 275 pounds, ranks just behind the tiger, the lion, and the jaguar in size. Although size is certainly the most obvious characteristic distinguishing the two groups (and is unduly emphasized by their labels), it is just one of the marks that define them.

Only one other cat, the clouded leopard, fails to fit in one of these two genera. Because this small Southeast Asian cat has several anatomical peculiarities characteristic of both *Felis* and *Panthera,* it is placed in own genus, *Neofelis.*

The following characteristics of each group highlight the evolutionary split that now separates the two, but, as stated, these characteristics provide few clues as to the manner in which this division occurred.

Hyoid Bone

Every great cat has an incompletely ossified hyoid bone supporting the tongue (connected to the base of the skull by extensions of cartilage), a structure that enables it to roar. In the lesser cats, their hyoid bones are completely ossified, which prevents them from roaring but enables them to make purring, mewing, yowling, or muttering sounds. The great cats can also purr but only during exhalation; lesser cats purr when exhaling, as well as inhaling. Exceptions from both groups include the snow leopard, which does not roar but can purr when exhaling and inhaling; the black-footed cat, which can purr immediately after birth but later develops a high-pitched roar; the cougar of the Andes, which makes a pumping roar; and the golden cat, which is said to sound like a leopard.

Pupils

The round shape of the pupils of the eyes of the great cats is often listed as one of their distinguishing marks; the lesser cats have elliptical pupils. This distinction is corroborated by the pupil's ability to retain its shape when its size adjusts to changing light levels. However, when a great cat's eyes are subjected to bright daylight, the pupils do become elliptical. Conversely, in dim light the pupils of some of the lesser cats, the fishing cat, for example, become round; and in bright light the cougar's pupils constrict to small discs.

Nose

The nose of a great cat has a covering of hair that reaches its front edge. On a lesser cat, the front ridge and top of the nose is hairless.

Claws

Except for the flat-headed cat, lesser cats retract their claws into sheaths that are shorter on the inner side of each paw; except for the cheetah, which has no protective sheaths at all, the sheaths protecting the claws of great cats are all equal in length.

Paws

At rest, great cats hold their paws out in front of themselves and extend their tails straight back. The lesser cats bend their paws at the wrists and tuck them under their chests, and they wrap their tails around to the front of their bodies.

Feeding

All great cats feed from a lying position, whereas lesser cats feed from a crouch.

Hare ● Rabbit

Is Bugs Bunny really a wascally wabbit or a harebrained hare in disguise? Considering his screen persona—a maverick personality mated to a lanky body with long ears—Bugs is probably a jackrabbit but no rabbit. Jackrabbits are hares, not rabbits. Neither is the snowshoe rabbit, whose alternative name, varying hare, better describes this winter-white and summer-brown hare. Others in the family carry names just as erroneous: the Belgian hare, the Hispid hare, and the red rock hare are all rabbits. But these "rabbit" and "hare" labels have not always been regarded as misnomers. It is only since zoologists began restricting their use to specific species that "rabbit" and "hare" took on narrower meanings. Originally, the two terms—"rabbit" borrowed from the Old French and "hare" from prehistoric West German languages—were simply synonyms referring to any member of the rabbit-hare family Leporidae.

When the Anglo-Saxons invaded Britain in the fifth century A.D., they brought with them the word *har*, meaning "gray." Taking *har* as a root word, the British created the words "herring" for certain gray fish and "hare" for many of Britain's grayish, rabbitlike mammals. Six centuries later, when the Normans invaded Britain, they introduced their word *robbe* for the same animal, which the British eventually changed to "rabbit." Only adopted in those localities where Norman influence was strongest, "rabbit" never fully replaced "hare" among the British. As a result, "rabbit" and "hare" eventually came to denote several different-looking creatures, but they implied no taxonomic difference.

However, as naturalists began studying rabbits and hares, two distinct forms began to emerge. One gave birth to naked, sightless, and toothless young; the other gave birth to those that were furry, were sighted, and bore teeth. Most in the first category had long been called rabbits, and many of the latter had been called hares. This led to the notion that despite the mixed use of "rabbit" and "hare," all those born naked, and so on, should be distinguished as rabbits, and only those born furred, and so on, should be called hares.

Although zoologists embrace this assignation, they have yet to endorse a restricted use of "rabbit" and "hare." But as a practical matter, most feel it appropriate to limit "hare" to those of the genus *Lepus* and to reserve "rabbit" for all others in the family. An exception is usually made for the Bushman hare of *Bunolagus*. This extremely rare animal is often linked to *Lepus* because of its resemblance to the jackrabbit—a creature so named because its long ears resemble those of a jackass.

Of the 24 species of rabbits in the world, those best known and described are the European rabbit *(Oryctolagus cuniculus)* and the 13 cottontails *(Sylvilgus* spp.). Accordingly, science generally distinguishes rabbits from hares based on the characteristics of these 14 rabbit species and the 21 hares of *Lepus*.

In popular literature, the features most often cited as distinguishing the two concern their birth. When born, rabbits are atrical; that is, they are naked, have closed eyes, have no teeth, and cannot walk. Hares, precocial animals, are born fully furred, with eyes open, with teeth, and can run about soon after birth. Much of this difference is attributed to their gestation periods: that of rabbits lasts only 26 to 30 days, whereas hares remain in the uterus from 36 to 47 days. The result of such births is that hares need give their young only scant attention, whereas rabbits must closely attend to their offspring for several weeks. In order for rabbits to do so, they build a fur-lined nest before giving birth. The European rabbit does this either in a burrow it constructs or in a communal den. With the exception of the pygmy rabbit, cottontails do not make burrows, nor do they form dens. These North American species choose to use burrows made by other animals or to build nests in sheltered underbrush. Most hares do not construct burrows either, although the snowshoe and the arctic hares do so for protection against their harsh environment. Neither do hares typically build nests but prefer giving birth in "forms," depressions scooped in the earth.

This contrast in nesting sites reflects a basic behavioral difference between hares and rabbits. Hares, well adapted for quick sprinting and hard running, favor open country with its unobstructed view of potential danger. Rabbits, able only to hop away from harm, stick to hilly, wooded, or overgrown locations, where they can quietly hide in foliage or behind broken terrain. This tends to make the rabbit a more social or communal animal than the hare. Hares, which range over a much larger territory, are typically solitary creatures.

Zoologists use other characteristics, besides the type of birth, to distinguish rabbits from hares. As adults, rabbits tend to be smaller than hares, their ears have no darkened tips, and their hind legs are shorter. Not as obvious, the rabbit has a longer palate; the joint between the interparietal and parietal bones of the skull is distinct; and because it does not engage in long-distance running, its muscles have shorter fibers and its flesh is light, rather than the dark red of the hare. In all, the rabbit is a more nimble animal, which, coupled with its acute vision, allows it to quickly and deftly maneuver amid brush. For the less well-sighted hare, its prowess lies in its speed and endurance.

Jaguar ● Cougar ● Leopard ● Panther

The jaguar is not only a good-size cat, it is the largest of all the New World cats. The average male weighs between 175 and 225 pounds, stands close to 30 inches high at the shoulder, and reaches 5 to 6 feet in length, excluding its tail. Its domain extends from the chaparral scrub of northern Mexico (the last sighting in the United States was made in 1971) down to the Patagonian plains and low-altitude brush of the Andes mountains of South America. While jaguars do well in these settings, most prefer living in wooded environments such as the Amazon rain forest and particularly where there is a stream or another source of water.

Besides providing a cooling swim, water sources are excellent places to find game. Caimans, turtles, fish, snakes, lizards, deer, capybaras, tapirs, and peccaries all figure into the jaguar's streamside menu. Jaguars also find trees a good place from which to prey. Reputed to be the best tree climber of all the big cats, a jaguar will pursue monkeys among the sturdier branches or will lie on a low limb and wait for a meal to pass below. Springing to attack, the jaguar easily takes down smaller quarry with a swipe of its claws. In attacking larger game, it sinks its teeth into its victim's cranium and crushes the skull. This is considerably different from the technique used by the other large cats, such as lions, tigers, and leopards, which kill by attacking the necks of their prey. Although the jaguar's method of attack is unusual, it is very efficient. Indians, recognizing its prowess in killing, named it *yaguara,* a word meaning "the animal that kills in one bound" and the derivation of "jaguar." Yet savage as it is, unlike the lion, the tiger, or the leopard, the jaguar is not known to be a man-eater. It has killed humans on occasion, but it never seems to have developed a particular liking for us.

The jaguar is a heavily built cat with a deep chest; a broad, heavy head; and short, powerful legs, very similar to the build of a small tiger. Early European explorers, noting this resemblance, named it *el tigre,* a word still used by Central and South American natives. Although the jaguar and the tiger may have a similar build, their different coats preclude confusing the two: the tiger is easily identified by its well-known stripes, whereas the jaguar's coat is dotted with dark blotches and rosettes. However, these rosettes often have led to its confusion with the leopard, a cat whose coat is almost identical. But "almost" is an important qualifier. Set against a color field ranging from light yellow to reddish brown, the rosettes marking the jaguar's flank run in a somewhat horizontally parallel pattern, with each rosette consisting of one or two spots surrounded by a ring of small blotches. Blotches on the legs, the head, and the tail form only token rings or simply single spots.

In contrast, the rosettes that mark the leopard's coat tend to occur more randomly and never enclose a spot. This difference in rosettes is probably the most frequently cited feature used in distinguishing the two animals. But the leopard's coat also differs from that of the jaguar's in its background coloring and length. Varying from a tawny or grayish yellow to chestnut brown, the fur of leopards living in warm climates is fairly short, whereas the fur of those living in colder areas is longer and silkier. Leopards living in open country also tend to be lighter in color than the more densely spotted forest-dwelling leopards. Occasionally, a leopard or a jaguar (rarely a cougar) will be born with a very dark, or even black, coat. Although such melanistic mutations retain their rosettes and spots, these markings may be difficult to discern. Not a common coloration among the leopards of Africa, this dark mutation is most often found in those living in India, Southeast Asia, and in particular the Malay Peninsula, where it is estimated to affect half the leopard population.

The leopard, a bit larger than the cheetah, is a smaller and shorter-tailed cat than the jaguar, yet it is still big enough to take down a good-size antelope. An extremely stealthy cat, the leopard steals as close as possible to its prey before pouncing. This tactic, which enables it to snare game that eludes other cats, allows the leopard to survive in a wide range of habitats. From arid savannas to humid tropical rain forests to cool mountain foothills, leopards have prevailed by dining on everything from dung beetles and cane rats to baboons and domestic cattle. Although leopards can catch most anything they want, most opt for a limited menu, which differs with each individual. Able to easily adapt to its surroundings, the leopard became the most widely distributed member of the big-cat group. While now found only in Southeast Asia and the southern two-thirds of Africa, in the past the leopard's range included the Middle and the Near East.

There is a large cat that does stand in for the leopard in the Middle and the Near East, but its preferred habitat is above the tree line of mountains (in altitudes ranging from 9,000 to 19,700 feet). This is the snow leopard of Siberia, Tibet, and Afghanistan, a cat well protected from the cold by its long, heavy, cream-colored coat. But the snow leopard is not a cold-climate variation of a leopard or even a true leopard.

Like the leopard and the jaguar, the snow leopard, or ounce, as it is also known, belongs to the lion and tiger genus *Panthera* and bears the full scientific name of *Panthera unica*—the leopard and the jaguar are classified as *P. pardus* and *P. onca,* respectively. This classification indicates that although the snow leopard and the leopard may be closely related, their evolutions took somewhat different paths. The snow leopard, a slightly smaller animal than the leopard, weighs 55 to 165 pounds and stands 24 inches at the shoulder, only a bit under the leopard's 62 to 198 pounds and 31-inch shoulder height. Because of its freezing habitat, the snow leopard wears an exceptionally heavy coat consisting of a dense underfur overlaid by fluffy guard hairs. Its tail is equally dense and unusually long—3 feet—almost as long as its head-to-rump length.

While there may be a shred of validity to the snow leopard's claim to the label "leopard," the clouded leopard, which makes its home in the rain forests of Southeast Asia, is decidedly a leopard in name only. Assigned its own genus, *Neofelis,* the clouded leopard is considered by some to be a "bridge" between the big cats of *Panthera,* the cheetah of *Acinonyx,* and the small cats of *Felis.* A long-bodied cat with cloudlike markings, blotches, and short stripes, the clouded leopard is noted for having the largest canine teeth, proportional to its size, of all living cats.

Although the clouded and the snow leopards may be ill served by the name "leopard," at least they have not been saddled with the enigmatic label "panther." Most commonly bestowed upon the leopard, "panther" is also a name sometimes given to the jaguar and the cougar. It is a name whose uncertain origin and questionable application have led to considerable confusion and speculation, particularly as to what, exactly, a panther is and how it differs, if at all, from a leopard.

In earlier times, when ancient Latin writers penned descriptions of the wildlife encountered by various travelers and explorers, much of the exotic game these adventurers came across was poorly known and understood. References to the animal some writers called a panther often stood in contrast to the descriptions of the same animal that other writers called a leopard. Consequently, it appeared that panthers and leopards were two distinct animals. However, "leopard" was constructed from the names of two animals thought to be its parents, the lion and the pard. Yet, as well as can be determined, a pard was simply a panther—described by one fourteenth-century writer as a panther with "white specks."

So, inasmuch as there is no such animal as a panther, distinct from the leopard, with which a lion could mate (all other cats having been accounted for), the presumption that the leopard was a third form of cat—a crossbreed—was erroneous. Some nineteenth-century scientists thought to distinguish panthers as large leopards, but this distinction never gained acceptance. Today, "panther" remains a vague synonym for "leopard" and is often used with the term "black" to describe the melanistic forms. "Black panther" has also been used to describe such jaguars but not the cougar.

The only cougar called a panther is the subspecies *Felis concolor coryi* of Florida, an animal that has never borne a black or even darkened coat— in only four cases have large black cats turned up in Florida, and these were melanistic bobcats. Sightings of black cougars within the United States have been reported, but no evidence supporting such claims has ever surfaced. However, black cougars do occur in Central and South America. Depending on their location, they are known as *pantera negra* (black puma), *cadupani* (black lion), and other variations on the term "black." But none of these labels—particularly, "black panther"—is thought to have made its way to Florida and given rise to the name "panther."

The first known use of "panther" as a name for the cougar was made in the seventeenth century by John Banister, a clergyman and a naturalist. "Panther," then as now, was a vague term and was used quite freely (as were "lion" and "tiger") for many of the newly discovered cats of the New World—in a 1695 South Carolina law regarding the killing of animals, the cougar was called a "tiger." Whether or not Banister was responsible for launching the use of "panther," the term soon became the cougar's established name throughout the eastern United States. In the backcountry the term was often corrupted into "painter," but this term, as well as "panther," eventually fell into disuse when the animal was killed off. The panther was eliminated from almost the entire eastern third of the country; the only place it has been able to survive is Florida and, on occasion, a few other southeastern states. Because the animal is nearing extinction, its old name "panther" will probably persist only as long as the cougar of the Southeast does.

The average male cougar, a bit smaller than the jaguar, weighs between 120 and 200 pounds and is about 7½ feet long. It has a slender build, a somewhat small head, and a long tail. Except for its light belly fur, the cougar's short coat is a solid color, which ranges from a reddish brown to tawny yellow to slate gray. Because it is a very shy cat—wildlife authorities do not consider the cougar to be particularly dangerous—it will normally go out of its way to avoid people. The only time it does come near humans is if it is starving and driven to attack livestock. Usually, however, a cougar can find enough deer, its main winter quarry, or rabbits, raccoons, squirrels, and other small wildlife to sustain itself.

Of the world's thirty cougar subspecies, thirteen live north of the Mexican border, with the balance forming an array of overlapping populations that extends to Tierra del Fuego. As a cat with an exceptionally large distribution, it is not surprising that "cougar" and "panther" are not its only names. Although the large number of subspecies indicates a considerable variation among cougars, there is no correspondence between such specific differences and its assortment of common names.

Even setting aside the many local names used by Central and South American Indians, the cougar's list of aliases is extensive. It must be noted that "cougar" is no more correct than are any of its other names; it is used here simply because of its broad use throughout the United States. Because many names overlap among the various cougar populations, no effort has been made to match them with specific geographies.

Cougar Aliases

American lion	leopard	panther
bay lion	lion	puma
black tiger	long-tailed cat	purple panther
brown ounce	Mexican lion	red lion
brown tiger	mountain devil	red ounce
California lion	mountain lion	red tiger
Catamount	mountain screamer	screamer
deer killer	painted devil	silver lion
deer tiger	painter	sneak cat
Florida panther	pampas cat	tiger
Indian devil	pampas lion	wild cat

Kangaroo ● Wallaby ● Wallaroo

In 1629, a Dutch sailor by the name of Francisco Pelsaert and his ship's complement of passengers became shipwrecked on a small island off the west coast of Australia. Eventually venturing onto the mainland, they were confronted by a curious collection of animal life and became the first Europeans to ever encounter a kangaroo. But whatever they may have called the animal, it is certain that it was not "kangaroo." "Kangaroo" entered the European vocabulary only after Pelsaert was long gone and then merely because of a supposed lack of communication. As the tale goes, the British sea captain James Cook (who rediscovered Australia in 1770) asked an Australian Aborigine what he called the strange animal, to which the native is supposed to have answered, *"Kangaroo,"* a word that meant "I don't understand [the question]." Cook, likewise not understanding (the answer), took it to be the animal's name. Another version of the word's origin suggests that "kangaroo" means "There he goes!"

Whatever the case, little care was taken in the subsequent use of "kangaroo," turning it into a broad term that now covers a wide variety of animals. Listed under the umbrella of "kangaroo" are the Monjon, a 2.5-pound cave-dwelling rock wallaby, and the Boongarry, a slightly larger animal that lives in trees. In all, there are between 48 and 95 species of marsupials cataloged as kangaroos (the difference in number reflects the ongoing dispute over their classification). Only a few are known simply as "kangaroo," most having been given additional names. Among these aliases are wallaby, wallaroo, pademelon, euro, flier, nail-tail, rat, scrubber, and stinker, plus scores of other Aboriginal names.[18]

Unfortunately, few of these common names were bestowed with any more care than was "kangaroo." The species *Macropus robustus* is widely known as the common wallaroo, but it also goes by the names hill kangaroo and euro. Few of the names really pinpoint specific forms (although they may denote specific species), and many names that once denoted fairly specific animals have since become quite broad. For example, true pademelons are small, compact versions of kangaroo that belong to the genus *Thylogale,* yet Australians frequently use "pademelon" for, among others, the Tammar wallaby of the *Macropus* genus. Even marsupials outside the kangaroo family Macropodidae (meaning "large foot") carry kangaroo names. The long-nosed potoroo, a small rodentlike member of the Potoroidae family, also goes by the names long-nosed rat-kangaroo and wallaby rat. Just as misnamed is the tiny one-pound musky rat-kangaroo of the same family.

As a matter of convenience, scientists split the marsupials of Macropodidae into four subgroups: (1) rock wallabies, (2) nail-tail wallabies, (3) New Guinea forest wallabies (Australia is not the kangaroo's only home), and (4) those making up the genera *Macropus* and *Wallabia*. *Macropus,* an extensive genus, includes the "typical" larger kangaroos, wallabies, and wallaroos (some authorities place wallaroos in the genus *Osphranter*). *Wallabia* has but a single species, the swamp wallaby *(W. bicolor),* an unusual kangaroo with unique genetic, anatomical, and behavioral characteristics.

But common to each member of Macropodidae are powerful hind limbs; short fore limbs; a long and robust tail; very similar dentition, with either absent or vestigial canines and a gap between the incisors and the molars; a low metabolic rate; an enlarged esophagus and stomach; and a small intestine in which fermentation breaks down the vast amounts of grass it consumes. Because kangaroos, wallabies, and wallaroos share so many characteristics, and inasmuch as their names have been liberally exchanged among its members, it is reasonable to question whether the three terms truly distinguish different forms. They do—but with qualification.

Although "kangaroo" has been bestowed upon quite a few species, as a popular label among Australians it is usually reserved for six specific members of *Macropus:* the western grey kangaroo, the eastern grey kangaroo, the red kangaroo, the common wallaroo, the black wallaroo, and the Antilopine wallaroo. Australians singled out these six because they are the largest species within the kangaroo family (weighing an average of 150 pounds), which has also earned them the title of "true kangaroos," although the term has no scientific standing.

The term "wallaroo" comes from the Aboriginal word *wolaru* and was used by native Australians for the larger kangaroo forms inhabiting the eastern region of New South Wales. The word was eventually picked up by European settlers, who initially used it only for the common wallaroo but later applied it to the Antilopine wallaroo and the black wallaroo. The common wallaroo and the Antilopine wallaroo do approach the size of the smallest kangaroo, weighing on average 105 pounds; however, the black wallaroo is much smaller, weighing only 48 pounds. Wallaroos, although smaller than the other three kangaroos, tend to be stockier and bear somewhat shaggier fur. Also, their muzzles are completely hairless, whereas those of the other kangaroos (and wallabies) are fully or at least partially haired. These nasal regions are also markedly enlarged in wallaroos, a result of their unusually broad third upper incisors.

Compared to the others of *Macropus,* most wallaroos have somewhat short and broad feet, with soles made up of thickened pads, protected by rough-textured skin. This allows them to easily clamber among the rocky terrain that most inhabit. Such terrain is the reason they are also called rock kangaroos. In fact, it is often said that one of the characteristics that distinguishes the wallaroos from the kangaroos is the rocky and mountainous regions they prefer. However, the Antilopine wallaroo, like the red kangaroo, favors grassy plains and savannas, and only occasionally roams rocky hill country. But the wallaroo is not the only kangaroo form running around rocky terrain; two genera of small wallabies, commonly called rock wallabies, also make their home in these areas.

The 7 wallabies of *Macropus* can best be distinguished from the others in the genus by their size. The 3 largest wallaby species weigh an average of 60 pounds, while the remaining 4 are considerably smaller, bringing the average weight of the 7 closer to 18 pounds. And this single feature, their size, seems to be the primary reason these animals are considered "wallabies." In his book *Kangaroos: The Marvelous Mob*, Terry Domico regards "wallaby" as a "loose term applied to any kangaroo species weighing less than 45 pounds (20 kilograms) as an adult."[19] This, of course, covers not only those of *Macropus* but all other wallabies as well: the rock, the nail-tail, and the forest wallabies. In *Walker's Mammals of the World, 6th ed.*, the wallabies of *Macropus* are said to be kangaroos weighing between 5 and 53 pounds.[20] The *New Larousse Encyclopedia of Animal Life* takes a bit different approach in defining a wallaby. It states that "wallaby" should be "reserved for the smaller species of kangaroo in which the feet are less than twenty-five centimeters [9.8 inches] long."[21] So, "wallaby" here embraces not only those of *Macropus*, which are simply smaller versions of the "true" kangaroo, but also those such as the spectacled hare-wallaby of *Lagorchestes*, which resembles a small rabbit.

Seal • Sea Lion • Walrus

Seals, sea lions, and walruses all belong to the order Pinnipedia, one of three mammal groups that left land millions of years ago for a life in the sea. Yet unlike the other two groups, the cetaceans (whales) and the sirenians (dugong and manatees), the pinnipeds never became fully aquatic. In order to give birth and suckle their young, they must return to land or climb onto ice. Also, cetaceans and sirenians possess large tails that serve as a primary means of propulsion, whereas the tiny tail of the pinniped is virtually useless. To propel themselves, pinnipeds use their flippers, whose form gave rise to their order's name, Pinnipedia, meaning "feather foot" or "wing foot."

Taxonomists divide Pinnipedia into two suborders (in some classification schemes these suborders are downgraded to superfamilies):

Order   Pinnipedia ("seals")
    Suborder   Phocidea (rear flippers point backward)
        Family   Phocidae ("true" seals)
    Suborder   Otarioidea (rear flippers turn forward)
        Family   Odobenidae (walrus)
        Family   Otariidae (eared seals)
            Subfamily   Otariinae (sea lions)
            Subfamily   Arctocephalinae (fur seals)

Phocidea, with its single family Phocidae, contains all the "true" seals; and Otarioidea, having two families, Odobenidae and Otariidae, contains, respectively, the walrus, and the fur seals and the sea lions. Altogether, there are 33 to 35 Pinniped species.

Despite this taxonomic classification, one will often find these mammals grouped under various combinations of their common names—regrettably, a confusing practice. Depending on the authority, the order also can be found divided into seals, walruses, and sea lions—the last also includes the fur seals. Or, true seals, walruses, and eared seals—the last includes sea lions. Or, sea lions, fur seals, walrus, northern phocids, and southern phocids. Moreover, the whole order is frequently called the seal order. With the pinnipeds organized under such loosely applied headings, it is not surprising that there may be questions as to what the terms "seal" and "sea lion" really refer to.

All 19 members of the family Phocidae are called seals. Often they are referred to as earless seals, hair seals, or true seals. Their designation as earless only means that they have no externally projecting ears. This is in contrast to the sea lion and the fur seal, which do have small ear appendages. Of the three labels, this is the only one that could be considered appropriate. The label "hair seal" is less acceptable. It is meaningful only when these seals are compared to the fur seals. To say that the phocoids are the "true" seals—more seal-like than any other pinniped—is imputing more status than warranted. Yet this is the most commonly used label, bestowed upon them as the first pinnipeds described by early

European naturalists and retained in part due to their theoretically once-unique origins.

Until 1989 it was fairly well established from fossil evidence that the earless ("true") seals descended from the mustelids: the weasels, the badgers, and so on. All other pinnipeds were thought to be linked to the ursids: the bears. This supported the conviction that inasmuch as the suborder Otarioidea was a mix of different pinnipeds, only some of which were called seals, the homogeneous suborder Phocoidea of only "seals" was unique and consequently merited the title of "true seals." However, DNA research, coupled with a close examination of pinniped flipper structure, now points to a common ancestor that lived 22 million years ago. The evidence indicates that during this era, the ancestors of the walrus, the fur seal, and the sea lion were swimming the seas, and that from these ancestral forms the "true" seals also evolved. So, whereas the phocoids are certainly unique in many respects, it would be a mistake to give the label "true" any real credence.

Within the suborder Otarioidea, the walrus stands apart from the sea lion and the fur seal. Never called a seal unless spoken of as a member of the "seal order," the walrus's shared features with others in Otarioidea include rear flippers that turn forward, flippers with naked soles, nails of varying size, and an exclusive marine habitat. No "true" seal shares these characteristics. In contrast to the sea lion and the fur seal, the walrus's testes are internal, its upper incisors are ungrooved, it has a large tympanic membrane in its ear, and there is no bony process on the skull above the eye socket; however, like the "true" seal, the walrus lacks external ears. Although these features do define the walrus, this pinniped is noted for its more apparent characteristics: a huge body (up to 2,700 pounds), very thick and wrinkled skin (over 2 inches thick in males), a broad mustached snout, and, of course, its two large ivory tusks. The walrus is also unusual in restricting itself to the arctic seas, often riding ice floes, whereas the other families include members who enjoy more temperate waters.

Finally, there is the Otariidae family, popularly called either eared seals or sea lions. While the term "eared seal" is appropriate—they are the only pinnipeds so visibly equipped—the label "sea lion" is misleading because not all members belong to the sea lion subfamily (Otariinae). The balance of the eared seals, which we commonly call fur seals, makes up the subfamily Arctocephalinae. Fur seals get their name from the many fine secondary hairs that accompany each larger hair, forming a dense mat that acts as an insulator. Other than the "true" seals, fur seals are the only other pinnipeds that zoologists customarily call "seals."

Sea lions, so named because their heavily furred necks resemble the manes of lions, are represented by one of the best-known pinnipeds: the California sea lion. Often seen in zoos, it is occasionally called a seal when it performs in circuses and marine shows. Compared to the fur seal, the sea lion has only token underhairs, its body is generally larger, the snout is broader, and there is one less rear upper canine. There are other small differences between the two, but neither these nor those just pointed out are particularly significant. In fact, some authorities feel that these differences do not justify dividing the family into subfamilies.

As an aid in telling the pinnipeds apart, it is helpful to note that "true" seals are unable to turn their rear flippers forward, and they swim by alternately stroking each of these flippers while keeping the fore flippers next to their bodies. On land, their rear flippers are useless for locomotion, so movement, provided by the fore flippers alone, is limited. The other pinnipeds, who can swing their rear flippers forward, use these as rudders when swimming and propel themselves with their fore flippers. Able to use all four flippers on land, they get around fairly well. As for sea lions, they are, in practicality, also seals (eared), very closely related to the fur seal. The walrus is a walrus, more closely related to the eared seals than to the "true" seals, but is seldom regarded as a seal.

Notes

Chapter 1: Nomenclature and Taxonomy

1. This is the Latinized, and most common, version of his given name, Karl Linnaeus, which he later changed to Karl von Linné, after being conferred nobility by the Swedish parliament.

2. Because few scientists today understand Latin grammar and many find its mandatory use unnecessarily burdensome, it may well be abandoned in the near future. Zoologists reviewed such a proposal in preparation of the third edition of the *International Code of Zoological Nomenclature*.

3. Louie Psihoyos and John Knoebber, *Hunting Dinosaurs* (New York: Random House, 1999), 105.

4. These particular endings are for plants other than algae or fungi.

5. The *International Code of Botanical Nomenclature* does not officially recognize these ranks.

6. As the foundation of cladistics, a scheme of investigation, this methodology is based on evolutionary systematics, an earlier form of analysis.

7. Haeckel's best-remembered legacy is probably the three-word proclamation (no longer accepted by any reputable biologist) that long bedeviled high school biology students: "Ontogeny recapitulates phylogeny."

8. As of 1993, "phylum" has been an optional designation for "division" in plant and fungi taxonomy.

9. A breed (for cultivar, substitute "plant" for "animal") can be defined as: "A group of animals that has been selected by man to possess a uniform appearance [or characteristic] that is inheritable and distinguishes it from other groups of animals within the same species" (Juliet Clutton-Brock, *Domesticated Animals from Early Times* [Austin: University of Texas Press, 1981], 26).

10. For a sobering look at the controversies that can arise within a particular field, simply examine the various cat family taxonomies that have been proposed in recent years.
11. Not all microbiologists feel that archaebacteria are deserving of this promotion.
12. Colin Tudge, *The Variety of Life* (New York: Oxford University Press, 2000).
13. In a recent reorganization of *Canis,* the domestic dog was reclassified as *Canis lupus familiaris,* a subspecies of wolf. However, because this is a somewhat new classification and has yet to be fully acknowledged, it has not been employed here.
14. Ernst Mayr, *The Growth of Biological Thought: Diversity, Evolution, and Inheritance* (Cambridge, Mass.: Belknap, 1982), 311.
15. Keith D. Rushford, *Conifers* (New York: Facts on File, 1987), 13.

Chapter 2: Plants

1. *Gurney's 2000 Spring Catalog* (Yankton, S. Dak.: Gurney's Seed and Nursery Co., 2000), 4.
2. Although most authorities recognize this ancestry of maize, there does remain some debate over its validity.
3. Today, "Indian corn," as used in the United States, refers to flint corn, a hard starch corn with multicolored kernels, which is often used for decoration.
4. Ansel Hartley Stubbs, *Wild Mushrooms of the Central Midwest* (Lawrence: University of Kansas Press, 1971), xvi.
5. Orson K. Miller, *Mushrooms of North America* (New York: Dutton, 1972), 358.
6. Kingsley Rowland Stern, *Introductory Plant Biology* (Dubuque, Iowa: Brown, 1994), 303.
7. Kent H. McKnight and Vera B. McKnight, *A Field Guide to Mushrooms: North America,* reissue ed. (Boston: Houghton Mifflin, 1998), 36.
8. Louis C. C. Krieger, *The Mushroom Handbook* (New York: Dover, 1967), 23.
9. Clyde M. Christensen, *Edible Mushrooms,* 2nd rev. ed. (Minneapolis: University of Minnesota Press, 1981), viii.
10. Stern, *Introductory Plant Biology,* 305.
11. Ibid.
12. John M. Kingsbury, *Poisonous Plants of the United States and Canada* (Englewood Cliffs, N.J.: Prentice-Hall, 1964), 88.
13. R. Gary Raham, *Dinosaurs in the Garden: An Evolutionary Guide to Backyard Biology* (Medford, N.J.: Plexus, 1988), 67.
14. Catherine Osgood Foster, *Plants-a-Plenty* (Emmaus, Pa.: Rodale, 1977), 271.
15. The use of "plant" in this discussion includes not only members of the Plant kingdom but also those of the Fungi kingdom.

16. In this discussion, the use of "spore" does not address organisms of the Bacteria or Protoctista kingdoms.
17. George W. B. Symonds, *The Shrub Identification Book* (New York: Morrow, 1963), 16.

Chapter 3: Terrestrial Invertebrates

1. Roger G. Bland, *How to Know the Insects,* 3rd ed. (Dubuque, Iowa: Brown, 1978), 372.
2. Bruce Ballenger, *The Lobster Almanac* (Chester, Conn.: Pequot, 1988), 64.

Chapter 4: Aquatic Life

1. Richard Ellis, *The Book of Sharks* (New York: Knopf, 1994), 16.
2. Ibid.
3. *Oxford English Dictionary,* Vol. 13 (1989), 241.

Chapter 5: Amphibians and Reptiles

1. Raymond Lee Ditmars, *Reptiles of the World: Crocodiles, Lizards, Snakes, Turtles and Tortoises of the Eastern and Western Hemispheres* (New York: Macmillan, 1933), 263.

Chapter 6: Birds

1. CNN broadcast, October 10, 1991.
2. Because of recent DNA evidence, some ornithologists now assign the New World vultures to the stork family, Ciconiidae.

Chapter 7: Mammals

1. Those of *E. m. inicus* can sometimes be found classified as *E. m. bengalensis* (Indian elephant) and *E. m. hirsutus* (Malaysian elephant).
2. Richard D. Estes, *The Behavior Guide to African Mammals: Including Hoofed Mammals, Carnivores, Primates* (Berkeley, Calif.: University of California Press, 1992), 8.
3. Some authorities consider the Przewalski horse to be more closely related to the extinct tarpan and classify it as *E. ferus przewalski.*
4. Not all zoologists agree on the makeup of *E. asinus:* some create the subgenus *Asinus* and assign the African ass to *E. Asinus africanus* and the donkey to *E. A. asinus.*

5. Jack Douglas Mitchell, *1992 Remington Calendar* (Wilmington, Del.: Remington Arms, 1992), 5.

6. Brown bear, American black bear, Asiatic black bear, polar bear, sun bear, sloth bear, spectacled bear, and giant panda.

7. The word also may have come from the Old English *grislic,* meaning "horrible," although this is doubtful.

8. Jeff Rennicke, *Bears of Alaskan Life and Legend* (Boulder, Colo.: Roberts Rinehart, 1987), 41.

9. Mike Cramond, *Of Bears and Man* (Norman: University of Oklahoma Press, 1986), 12.

10. Maurice Burton, ed., *The New Larousse Encyclopedia of Animal Life* (New York: Bonanza, 1981), 593.

11. Maria Pia Mannucci and Alessandro Minelli, *Great Book of the Animal Kingdom* (New York: Arch Cape, 1988), 329.

12. This is the domestic dog, one of eight species that make up the wolf/coyote genus *Canis.* Although the term "dog" is often used to indicate any or all members of the canid family, here it is used to denote only the domestic dog, *Canis familiaris.*

13. Recently, the American Society of Mammalogists reclassified the domestic dog as *Canis lupus familiaris,* a subspecies of wolf, and the wolf as *Canis lupus nubilus* and *occidentalis,* two subspecies. However, because this reclassification has yet to be fully acknowledged, it is not employed here.

14. Linehan, Edward J. "The Trouble with Dolphins." *National Geographic* (April 1979): 514.

15. Douglas M. Considine, ed., *Van Nostrand's Scientific Encyclopedia,* 8th ed. (New York: Van Nostrand Reinhold, 1995), 3324.

16. Ibid., 3323.

17. For a brief recap of these controversies, see R. M. Nowak, *Walker's Mammals of the World,* 5th ed., Vol. 2.

18. The 1988 report by the Australian Senate Select Committee on Animal Welfare observed, "The term 'kangaroo' is sometimes used to refer to all 48 species of [superfamily] Macropodoida but, at other times, it is used to refer to the six species in the macropod genus *Macropus* or the three largest species of this genus."

19. Terry Domico, *Kangaroos: The Marvelous Mob* (New York: Facts on File, 1993), 7.

20. Ronald M. Nowak, *Walker's Mammals of the World,* 6th ed. (Baltimore: Johns Hopkins University Press, 1999), 103.

21. Burton, *The New Larousse Encyclopedia of Animal Life,* 487.

Bibliography

Nomenclature and Taxonomy

Alexander, Peter, et al. *Biology: The Living World*. Englewood Cliffs, N.J.: Prentice-Hall, 1989.

Arnett, Ross H., and Richard L. Jaques Jr. *Insect Life: A Field Entomology Manual for the Amateur Naturalist*. Englewood Cliffs, N.J.: Prentice-Hall, 1985.

Augee, Michael, and Brett Gooden. *Echidnas of Australia and New Guinea*. Kingston, N.S.W., Australia: New South Wales University Press, 1995.

Bailey, Liberty Hyde, and Ethel Zoe Bailey. *Hortus Third: A Concise Dictionary of Plants in the United States and Canada*. New York: Macmillan, 1976.

Bakker, Robert T. *The Dinosaur Heresies: New Theories Unlocking the Mystery of the Dinosaurs and Their Extinction*. New York: Kensington, 1996.

Brooks, Daniel R., and Deborah McLennan. *Phylogeny, Ecology, and Behavior: A Research Program in Comparative Biology*. Chicago: University of Chicago Press, 1991.

Buschbaum, Ralph, et al. *Animals Without Backbones*, 3rd ed. Chicago: University of Chicago Press, 1987.

Clutton-Brock, Juliet. *Domesticated Animals from Early Times*. Austin: University of Texas Press, 1981.

Cogger, Harold G., ed. *Reptiles and Amphibians*. New York: Smithmark, 1992.

Considine, Douglas M., ed. *Van Nostrand's Scientific Encyclopedia*, 8th ed. New York: Van Nostrand Reinhold, 1995.

Curtis, Helena. *Biology*, 5th ed. New York: Worth, 1989.

Darton, Mike, and John Clark. *The Macmillan Dictionary of Measurement*. New York: Macmillan, 1994.

Dodson, Edward O., and Peter Dodson. *Evolution: Process and Product*, 3rd ed. Boston: PWS, 1985.

Gaffney, Eugene S., Lowell Dingus, and Miranda K. Smith. "Why Cladistics?" *Natural History* (June 1995).

Gamlin, Linda, and Gail Vines, eds. *The Evolution of Life.* New York: Oxford University Press, 1987.

Gotch, A. F. *Mammals, Their Latin Names Explained: A Guide to Animal Classification.* New York: Blandford, 1979.

———. *Latin Names Explained: A Guide to the Scientific Classification of Reptiles, Birds and Mammals.* New York: Facts on File, 1996.

Gould, Stephen Jay. "What Is a Species?" *Discover* (December 1992).

Greuter, Werner, et al. *International Code of Botanical Nomenclature* (Tokyo Code). Koenigstein, Germany: Koeltz Scientific Books, 1994.

Harris, Leon C. *Concepts in Zoology,* 2nd ed. Scranton, Pa.: HarperCollins, 1992.

Harris, Marvin. *Culture, People, Nature: An Introduction to General Anthropology,* 6th ed. New York: Addison-Wesley, 1997.

Holmes, Sandra. *Outline of Plant Classification.* New York: Longman, 1983.

Ingmanson, Dale E., and William J. Wallace. *Oceanography, an Introduction.* Belmont, Calif.: Wadsworth, 1995.

International Trust for Zoological Nomenclature 1985. *International Code of Zoological Nomenclature,* 3rd ed. Berkeley: University of California Press, 1985.

Jeffrey, Charles. *An Introduction to Plant Taxonomy,* 2nd ed. Cambridge, England: Cambridge University Press, 1982.

Keaton, William, et al. *Biological Science,* 4th ed. New York: Norton, 1986.

Krieg, Noel R., ed. *Bergey's Manual of Determinative Systematic Bacteriology.* Baltimore: Williams and Wilkins, 1994.

Margulis, Lynn, and Dorion Sagan. *What Is Life?* New York: Simon and Schuster, 1995.

Margulis, Lynn, and Karlene V. Schwartz. *Five Kingdoms: An Illustrated Guide to the Phyla of Life on Earth,* 3rd ed. New York: Freeman, 1997.

Mayr, Ernst. *The Growth of Biological Thought: Diversity, Evolution, and Inheritance.* Cambridge, Mass.: Belknap, 1982.

———. *This Is Biology: The Science of the Living World.* Cambridge, Mass.: Belknap, 1997.

McKnight, Tom L. *Physical Geography: A Landscape Appreciation.* Englewood Cliffs, N.J.: Prentice-Hall, 1998.

Miller, Stephan A., and John P. Harley. *Zoology: The Animal Kingdom.* Dubuque, Iowa: Brown, 1996.

Moore, David M. *The Story of Plant Life on Earth.* New York: Cambridge University Press, 1982.

Nadakavukaren, Mathew J., and Derek McCracken. *Botany: An Introduction to Plant Biology.* St. Paul, Minn.: West, 1985.

Nowak, Ronald M. *Walker's Mammals of the World,* 6th ed. Baltimore: Johns Hopkins University Press, 1999.

Parker, Sybil P. *Grzimek's Encyclopedia of Animals.* New York: McGraw-Hill, 1990.

Psihoyos, Louie, and John Knoebber. *Hunting Dinosaurs.* New York: Random House, 1999.

Ridley, Mark. *The Problems of Evolution.* New York: Oxford University Press, 1985.

Rushford, Keith D. *Conifers.* New York: Facts on File, 1987.

Sagan, Dorion, and Lynn Margulis. *Garden of Microbial Delights: A Practical Guide to the Subvisible World.* Boston: Harcourt Brace Jovanovich, 1995.

Stanley, Steven M. *Earth and Life Through Time.* New York: Freeman, 1986.

Stringer, Chris. *In Search of the Neanderthals: Solving the Puzzle of Human Origins.* New York: Thames and Hudson, 1995.

Tudge, Colin. *The Variety of Life.* New York: Oxford University Press, 2000.

Wyman, Donald, ed. *Wyman's Gardening Encyclopedia.* New York: Macmillan, 1987.

Young, J. Z. *The Life of Vertebrates,* 3rd ed. New York: Oxford University Press, 1991.

Plants

Alcamo, I. Edward. *Fundamentals of Microbiology,* 4th ed. Redwood City, Calif.: Benjamin/Cummings, 1994.

Alexander, Peter, et al. *Biology: The Living World.* Englewood Cliffs, N.J.: Prentice-Hall, 1989.

Allaby, Michael. *Dictionary of the Environment,* 3rd ed. New York: New York University Press, 1989.

———. *The Concise Oxford Dictionary of Botany.* Oxford, England: Oxford University Press, 1992.

Anderson, David A., and I. I. Holland. *Forests and Forestry,* 3rd ed. Danville, Ill.: Interstate, 1997.

Arora, David. *Mushrooms Demystified: A Comprehensive Guide to the Fleshy Fungi,* 2nd ed. Berkeley, Calif.: Ten Speed, 1986.

Barnhart, Robert K., ed. *Barnhart Dictionary of Etymology.* New York: Eilson, 1999.

Boyd, Robert F. *General Microbiology,* 2nd ed. St. Louis: Times Mirror/Mosby, 1988.

Bradley, Fern Marshall, and Barbra W. Ellis, eds. *Rodale's All-New Encyclopedia of Organic Gardening: The Indispensable Resource for Every Gardener.* Emmaus, Pa.: Rodale, 1992.

Brady, George S., and Henry R. Clauser. *Materials Handbook,* 14th ed. New York: McGraw-Hill, 1996.

Britton, Nathaniel Lord, and Addison Brown. *Illustrated Flora of the Northern United States and Canada,* Vol. 2. New York: Dover, 1999.

Brookes, Brian S. *The British Naturalists' Association Guide to Mountain and Moorland.* Marlborough, England: Crowood Press, 1985.

Christensen, Clyde M. *Edible Mushrooms,* 2nd ed. rev. Minneapolis: University of Minnesota Press, 1981.

Considine, Douglas M. *Van Nostrand's Scientific Encyclopedia,* 8th ed. New York: Van Nostrand Reinhold, 1995.

Coombs, Allen J. *The Complete Book of Shrubs.* Pleasantville, N.Y.: Reader's Digest, 1988.

Crockett, James Underwood. *Vegetables and Fruits.* New York: Henry Holt, 1986.

Elias, Thomas S. *Complete Trees of North America: A Field Guide and Natural History.* New York: Van Nostrand Reinhold, 1980.

Encyclopedia Americana. Danbury, Conn.: Grolier, 1997.

Rodale, J. I., ed. *Encyclopedia of Organic Gardening.* Emmaus, Pa.: Rodale, 2000.

Fortin, François. *The Visual Food Encyclopedia.* New York: Macmillan, 1996.

Foster, Catherine Osgood. *Plants-a-Plenty.* Emmaus, Pa.: Rodale, 1977.

Freethy, Ron. *The Making of the British Countryside.* North Pomfret, Vt.: David and Charles, 1981.

Fussell, Betty Harper. *The Story of Corn.* New York: Knopf, 1992.

Galle, Fred C. *Evergreen Shrubs.* New York: Time-Life Books, 1989.

Gamlin, Linda, and Gail Vines, eds. *The Evolution of Life.* New York: Oxford University Press, 1987.

Grant, Michael C. "The Trembling Giant." *Discover* (October 1993).

Graves, Arthur Harmon. *Illustrated Guide to Trees and Shrubs: A Handbook of the Woody Plants of the Northeastern United States and Adjacent Regions,* rev. ed. New York: Dover, 1992.

Gray, Asa. *Gray's Manual of Botany: A Handbook of the Flowering Plants and Ferns of the Central and Northeastern United States and Adjacent Canada.* New York: Van Nostrand, 1970.

Gurney's 2000 Spring Catalog. Yankton, S. Dak.: Gurney's Seed and Nursery Co., 2000.

Harris, Marvin. *Culture, People, Nature: An Introduction to General Anthropology.* New York: Addison-Wesley, 1997.

Haygreen, John G., and Jim L. Bowyer. *Forest Products and Wood Science: An Introduction.* Ames: Iowa State University Press, 1996.

Heiser Jr., Charles B. *Seed to Civilization: The Story of Food,* 3rd ed. Cambridge, Mass.: Harvard University Press, 1990.

Hewes, Jeremy John. *Redwoods: The World's Largest Trees.* New York: Smithmark, 1995.

Huxley, Anthony. *Green Inheritance: The World Wildlife Fund Book of Plants.* Village Station, N.Y.: Four Walls Eight Windows, 1992.

Jacobs, Louis L. *Quest for the African Dinosaur: Ancient Roots of the Modern World.* New York: Villard, 1993.

Johnson, Hugh. *Hugh Johnson's Encyclopedia of Trees,* 2nd ed. New York: Gallery, 1984.

Jonas, Gerald. *The Living Earth Book of North American Trees.* Pleasantville, N.Y.: Reader's Digest, 1993.

Keaton, William, et al. *Biological Science,* 4th ed. New York: Norton, 1986.

Kiester Jr., Edwin. "Prophets of Gloom." *Discover* (November, 1991).

Kingsbury, John M. *Poisonous Plants of the United States and Canada.* Englewood Cliffs, N.J.: Prentice-Hall, 1964.

Kirk, Ruth. *The Olympic Rainforest.* Seattle: University of Washington Press, 1992.

Krieger, Louis C. C. *The Mushroom Handbook.* New York: Dover, 1967.

Lawrence, Eleanor. *Henderson's Dictionary of Biological Terms.* New York: Wiley, 1989.

Lessard, W. O. *The Complete Book of Bananas.* United States: W. O. Lessard, 1992.

Magic and Medicine of Plants. Pleasantville, N.Y.: Reader's Digest, 1986.

Marsh, William M. *Earthscape: A Physical Geography.* New York: Wiley, 1987.

Masefield, G. B., et al. *The Oxford Book of Food Plants.* London: Oxford University Press, 1969.

Miller, Orson K. *Mushrooms of North America.* New York: Dutton, 1972.

McKenny, Margaret. *The Savory New Mushroom.* Seattle: University of Washington Press, 1987.

McKnight, Kent H., and Vera B. McKnight. *A Field Guide to Mushrooms: North America,* reissue ed. Boston: Houghton Mifflin, 1998.

Montagné, Prosper. *Larousse Gastronomique.* New York: Crown, 1988.

Moore-Landecker, Elisabeth. *Fundamentals of the Fungi,* 4th ed. Englewood Cliffs, N.J.: Prentice-Hall, 1996.

Morgan, Adrian. *Toads and Toadstools.* Berkeley, Calif.: Celestial Arts, 1995.

Nadakavukaren, Mathew J. and Derek McCracken. *Botany: An Introduction to Plant Biology.* St. Paul, Minn.: West, 1985.

Newcomb, Duane G., and Karen Newcomb. *The Complete Vegetable Gardener's Source Book,* rev. ed. Englewood Cliffs, N.J.: Prentice-Hall, 1989.

Page, Jake, and the editors of Time-Life Books. *The Forest.* New York: Time-Life Books, 1987.

Peirce, Lincoln C. *Vegetables: Characteristics, Production and Marketing.* New York: Wiley, 1987.

Pielou, E. C. *The World of Northern Evergreens.* Ithaca, N.Y.: Comstock, 1988.

Polumin, Oleg, and Martin Walters. *Guide to the Vegetation of Britain and Europe.* New York: Oxford University Press. 1985.

Prescott, G. W. *How to Know Aquatic Plants,* 2nd ed. Dubuque, Iowa: Brown, 1980.

Preston, Richard Joseph. *North American Trees (Exclusive of Mexico and Tropical United States),* 4th ed. Ames: Iowa State University Press, 1989.

Raham, R. Gary. *Dinosaurs in the Garden: An Evolutionary Guide to Backyard Biology.* Medford, N.J.: Plexus, 1988.

Raven, Peter H., and George B. Johnson. *Biology,* 4th ed. Dubuque, Iowa: Brown, 1996.

Riemer, Donald H. *Introduction to Freshwater Vegetation.* Malabar, Fla.: Krieger, 1993.

Rushford, Keith D. *Conifers.* New York: Facts on File, 1987.

Sharpe, Grant Williams, and others. *Introduction to Forestry,* 6th ed. New York: McGraw-Hill, 1994.

Shigo, Alex. *A New Tree Biology Dictionary.* Durham, N.H.: Shigo and Trees, 1986.

Sicherman, Al. "Shallots." *Star Tribune* (Minneapolis), February 20, 1991.

Smith, Alexander, and Nancy Smith Weber. *The Mushroom Hunter's Field Guide.* Ann Arbor, Mich.: Thunder Bay, 1996.

Snyder, Leon C. *Trees and Shrubs for Northern Gardens: New Revised Edition.* Minneapolis: University of Minnesota Press, 2000.

Splittstoesser, Walter E. *Vegetable Growing Handbook,* 3rd ed. New York: Kluwer Academic, 1990.

Starr, Cecie, and Ralph Taggart. *Biology: The Unity and Diversity of Life,* 6th ed. Belmont, Calif.: Wadsworth, 1992.

Stern, Kingsley Rowland. *Introductory Plant Biology.* Dubuque, Iowa: Brown, 1994.

Stoddard, Charles Hatch, and Glenn M. Stoddard. *Essentials of Forestry Practice,* 4th ed. New York: Wiley, 1987.

Strahler, Alan H., and Arthur N. Strahler. *Modern Physical Geography,* 4th ed. New York: Wiley, 1991.

Stubbs, Ansel Hartley. *Wild Mushrooms of the Central Midwest.* Lawrence: University of Kansas Press, 1971.

Swindells, Philip. *At the Water's Edge: Gardening with Moisture-Loving Plants.* London: Ward Lock, 1991.

Symonds, George W. B. *The Shrub Identification Book.* New York: Morrow, 1963.

Taylor, Norman. *Taylor's Guide to Trees.* Boston: Houghton Mifflin, 1988.

———. *Taylor's Guide to Vegetables and Herbs.* Boston: Houghton Mifflin, 1987.

United States Agricultural Research Services. *Common Weeds of the United States.* New York: Dover, 1987.

Vitt, Dale H. *Mosses, Lichens and Ferns of Northwest North America.* Seattle: Lone Pine, 1993.

Wyman, Donald, ed. *Wyman's Gardening Encyclopedia.* New York: Simon and Schuster, 1987.

Young, Raymond A., and Ronald L. Giese, eds. *Introduction to Forest Science,* 2nd ed. New York: Wiley, 1990.

Terrestrial Invertebrates

Arnett, Ross H., and Richard L. Jaques Jr. *Insect Life: A Field Entomology Manual for the Amateur Naturalist.* Englewood Cliffs, N.J.: Prentice-Hall, 1985.

Ballenger, Bruce. *The Lobster Almanac.* Chester, Conn.: Pequot, 1988.

Banister, Keith, ed. *The Encyclopedia of Aquatic Life.* New York: Facts on File, 1985.

Barnes, Robert D. *Invertebrate Zoology,* 6th ed. Philadelphia: Saunders, 1994.

Bland, Roger G. *How to Know the Insects,* 3rd ed. Dubuque, Iowa: Brown, 1978.

Bornancin, B. *World of Insects.* Memphis, Tenn.: Burke, 1983.

Borror, Donald J., and Richard E. White. *Peterson Field Guide to the Insects of America North of Mexico.* Boston: Houghton Mifflin, 1998.

Brodie, Edmund. *Venomous Animals.* New York: Golden, 2000.

Brusca, Richard C. *Invertebrates.* Sunderland, Mass.: Sinauer Associates, 1990.

Bull, Bartle. *Safari: A Chronicle of Adventure.* New York: Penguin, 1992.

Burton, Maurice, ed. *The New Larousse Encyclopedia of Animal Life.* New York: Bonanza, 1981.

Buschbaum, Ralph, et al. *Animals Without Backbones,* 3rd ed. Chicago: University of Chicago Press, 1987.

Chinery, Michael. *Dictionary of Animals.* New York: Arco, 1984.

Considine, Douglas M., ed. *Van Nostrand's Scientific Encyclopedia,* 8th ed. New York: Van Nostrand Reinhold, 1995.

Crompton, John. *The Hunting Wasp.* New York: Lyons, 1987.

Feltwell, John. *The Encyclopedia of Butterflies.* New York: Prentice-Hall, 1993.

Headstrom, Richard. *Lobsters, Crabs, Shrimps, and Their Relatives.* South Brunswick, N.J.: Barnes, 1979.

Helfer, Jacques R. *How to Know Grasshoppers, Crickets, Cockroaches and Their Allies,* 3rd ed. New York: Dover, 1987.

Henrich, Bernd. *Bumble Bee Economics.* Cambridge, Mass.: Harvard University Press, 1979.

Holldobler, Bert. *The Ants.* Cambridge, Mass.: Harvard University Press, 1990.

Hutchins, Ross E. *Grasshoppers and Their Kin.* New York: Dodd, Mead, 1972.

Ingmanson, Dale E., and William J. Wallace. *Oceanography: An Introduction.* Belmont, Calif.: Wadsworth, 1995.

Johnson, Jinny. *Simon and Schuster Guide to Insects and Spiders*. New York: Simon and Schuster, 1996.

Klausnitzer, Bernhard. *Insects: Their Biology and Cultural History*. New York: Universe Books, 1987.

Lawrence, Eleanor. *Henderson's Dictionary of Biological Terms*, 11th ed. New York: Wiley, 1995.

Lutz, Frank Eugene. *Field Book of Insects*. New York: Putnam, 1984.

McGavin, George C., and Richard Lewington. *American Nature Guides: Insects*. New York: Smithmark, 1992.

Milne, Lorus. *Audubon Society Field Guide to North American Insects and Spiders*. New York: Knopf, 1980.

Oldroyd, Harold. *Natural History of Flies*. New York: Norton, 1964.

Opler, Paul A. *Peterson's First Guide to Butterflies and Moths*. Boston: Houghton Mifflin, 1994.

O'Toole, Christopher. *Encyclopedia of Insects*. New York: Checkmark, 1995.

Parker, Sybil P. *Grzimek's Encyclopedia of Animals*. New York: McGraw-Hill, 1990.

Pearse, Vicki. *Living Invertebrates*. Palo Alto, Calif.: Blackwell Scientific, 1987.

Preston-Mafham, Ken. *Grasshoppers and Mantids of the World*. New York: Sterling, 1998.

———, and Ron Preston-Mafham. *Spiders of the World*. New York: Facts on File, 1984.

Pyle, Robert Michael. *The Audubon Society Handbook for Butterfly Watchers*. New York: Scribner's, 1984.

Raham, R. Gary. *Dinosaurs in the Garden: An Evolutionary Guide to Backyard Biology*. Medford, N.J.: Plexus, 1988.

Scheffel, Richard L. *Reader's Digest ABC's of Nature: A Family Answer Book*. Pleasantville, N.Y.: Reader's Digest, 1984.

Scott, James A. *The Butterflies of North America: A Natural History and Field Guide*. Stanford, Calif.: Stanford University Press, 1992.

Swan, Lester A., and Charles S. Papp. *The Common Insects of North America*. New York: Harper and Row, 1972.

Wild Wild World of Insects and Spiders. New York: Time-Life Films, 1977.

Wootton, Anthony. *Insects of the World*. New York: Sterling, 1998.

Zahradnik, J. *Bees, Wasps and Ants*. London: Hamlyn, 1991.

Aquatic Life

Abbott, Robert Tucker. *Kingdom of the Sea Shell*. Melbourne, Fla.: American Malacologists, 1993.

Allen, Thomas B. *The Shark Almanac.* New York: Lyons, 1999.

Arseniev, V. A. *Atlas of Marine Mammals.* Neptune City, N.J.: T. F. H., 1986.

Ayto, John. *Dictionary of Word Origins.* New York: Arcade, 1993.

Banister, Keith, ed. *The Encyclopedia of Aquatic Life.* New York: Facts on File, 1985.

Barnes, Robert D. *Invertebrate Zoology,* 6th ed. Philadelphia: Saunders, 1994.

Barnhart, Robert K., ed. *Barnhart Dictionary of Etymology.* New York: Eilson, 1988.

Bauer, Erwin A. *Saltwater Fisherman's Bible,* 3rd ed. rev. Garden City, N.Y.: Doubleday, 1991.

Bauman, Paul C., and James Jager. *Freshwater Fisherman's Companion.* New York: Van Nostrand Reinhold, 1983.

Boschung, Herbert T., et al. *Audubon Society Field Guide to Fishes, Whales, and Dolphins.* New York: Knopf, 1983.

Burton, Maurice, ed. *The New Larousse Encyclopedia of Animal Life.* New York: Bonanza, 1981.

Burton, Robert. *Wonders of the Sea.* New York: Exeter, 1983.

Buschbaum, Ralph, et al. *Animals Without Backbones,* 3rd ed. Chicago: University of Chicago Press, 1987.

Castro, Jose. *The Sharks of North American Waters.* College Station: Texas A & M University Press, 1996.

Chinery, Michael. *Dictionary of Animals.* New York: Arco, 1984.

Considine, Douglas M., ed. *Van Nostrand's Scientific Encyclopedia,* 8th ed. New York: Van Nostrand Reinhold, 1995.

Cook, Joseph J. *The Nocturnal World of the Lobster.* New York: Dodd, Mead, 1971.

Cousteau, Jacques-Yves, and Philippe Diole. *Octopus and Squid, the Soft Intelligence.* Garden City, N.Y.: Doubleday, 1973.

Davis, Richard A. *Oceanography: An Introduction to Marine Environments,* 2nd ed. Dubuque, Iowa: Brown, 1991.

Eddy, Samuel. *How to Know Freshwater Fishes,* 2nd ed. New York: McGraw-Hill, 1978.

———, and James C. Underhill. *Northern Fishes,* 3rd ed., rev. expd. Minneapolis: University of Minnesota Press, 1974.

Ellis, Richard. *Monsters of the Sea.* New York: Knopf, 1994.

———. *The Book of Sharks.* New York: Knopf, 1994.

Eschmeyer, William N., and Earl S. Herald. *Peterson Field Guide to Pacific Coast Fishes of North America: From the Gulf of Alaska to Baja California.* Boston: Houghton Mifflin, 1983.

Ferguson, Ava, and Gregor Calliet. *Sharks and Rays of the Pacific Coast.* Monterey, Calif.: Monterey Bay Aquarium, 1990.

Fitzpatrick, Joseph F. *How to Know Freshwater Crustaceans.* Dubuque, Iowa: Brown, 1983.

"Fixit." *Star Tribune* (Minneapolis), June 9, 1992.

Fortin, François, ed. *The Visual Food Encyclopedia.* New York: Macmillan, 1996.

Goodson, Gar. *Fishes of the Pacific Coast: Alaska to Peru, Including the Gulf of California and the Galapagos Islands.* Stanford, Calif.: Stanford University Press, 1988.

Headstrom, Richard. *All About Lobsters, Crabs, Shrimps, and Their Relatives.* New York: Dover, 1985.

Hedeen, Robert A. *The Oyster: The Life and Lore of the Celebrated Bivalve.* Centerville, Md.: (Tidewater) Cornell Maritime, 1986.

Ingmanson, Dale E., and Wallace J. Wallace. *Oceanography: An Introduction.* Belmont, Calif.: Wadsworth, 1995.

Lawrence, Eleanor. *Henderson's Dictionary of Biological Terms.* New York: Wiley, 1989.

Meffe, Gary K., and Franklin F. S. Nelson Jr. *Ecology and Evolution of Live-bearing Fishes.* Englewood Cliffs, N.J.: Prentice Hall, 1989.

Meglitsh, Paul A., and Frederick R. Schram. *Invertebrate Zoology,* 3rd ed. New York: Oxford University Press, 1991.

Meinkoth, Norman A. *Audubon Society Field Guide to North American Seashore Creatures.* New York: Knopf, 1981.

Moss, Sanford A. *Sharks: An Introduction for the Amateur Naturalist.* Englewood Cliffs, N.J.: Prentice-Hall, 1984.

Nelson, Joseph S. *Fishes of the World,* 3rd ed. New York: Wiley, 1994.

Parker, Sybil P. *Grzimek's Encyclopedia of Animals.* New York: McGraw-Hill, 1990.

Pearse, Vicki. *Living Invertebrates.* Palo Alto, Calif.: Blackwell Scientific, 1987.

Reader's Digest Association. *Sharks: Silent Hunters of the Deep.* Pleasantville, N.Y.: Reader's Digest, 1986.

Reynolds, Clarence V. "Warm Blood for Cold Water." *Discover* (January 1994).

Roberts, Mervin F. *Pearlmakers: The Tidemarsh Guide to Finding and Gathering Clams, Oysters, Mussels, and Scallops.* Dallas: Saybrook, 1984.

Robins, Richard, C. *The Freshwater Fish of the World.* Boston: Houghton Mifflin, 1986.

Rupert, Edward. *Seashore Animals of the Southeast: A Guide to Common Shallow-Water Invertebrates of the Southeastern Atlantic Coast.* Columbia: University of South Carolina Press, 1988.

Schur, Norman W. *British English A to Zed.* New York: Facts on File, 2001.

Seff, Philip, and Nancy Seff. *Our Fascinating Earth,* rev. ed. Chicago: Contemporary, 1996.

Simpson, J. A., and E. S. C. Weiner. *Oxford English Dictionary,* 2nd ed., 20 Vols. New York: Oxford University Press, 1989.

Springer, Victor G. *Sharks in Question: The Smithsonian Answer Book.* Washington, D. C.: Smithsonian Institution, 1989.

Stannard, Herald. *Fishes of North America.* Garden City, N.Y.: Doubleday, 1972.

Stevens, Jane E. "Life on a Melting Continent." *Discover* (August 1995).

Sweeny, James B. *A Pictorial History of Sea Monsters and Other Dangerous Marine Life.* New York: Crown, 1972.

The Fish List: F.D.A. Guide to Acceptable Market Names for Food Fish Sold in Interstate Commerce 1988. Washington, D.C.: U. S. Government, 1988.

The Seafood List. [web site: http://www.cfsan.fda.gov/~frf/seaintro.html]. U.S. Food and Drug Administration, 2001.

Varasdi, J. Allen. *Myth Information.* New York: Ballantine, 1996.

Walls, Jerry G. *Encyclopedia of Marine Invertebrates.* London: T. F. H., 1982.

Wheeler, Alwyne. *Fishes of the World: An Illustrated Dictionary.* New York: Macmillan, 1975.

Whitehead, Peter. *How Fishes Live.* New York: Elsevier, 1975.

Whitfield, Philip, ed. *Macmillan Illustrated Animal Encyclopedia.* New York: Macmillan, 1999.

World Record Game Fishes. Pompano Beach, Fla.: International Game Fish Association, 1991.

Wye, Kenneth R. *The Encyclopedia of Shells.* New York: Facts on File, 1991.

Amphibians and Reptiles

Ballinger, Royce E. *How to Know the Amphibians and Reptiles.* Dubuque, Iowa: Brown, 1983.

Bauchot, Roland, ed. *Snakes: A Natural History.* New York: Sterling, 1994.

Burton, Maurice, ed. *The New Larousse Encyclopedia of Animal Life.* New York: Bonanza, 1981.

Carr, Archie. "Alligators: Dragons in Distress." *National Geographic* (January 1967).

Collins, Mark, ed. *The Last Rainforest: A World Conservation Atlas.* New York: Oxford University Press, 1990.

Conant, Roger, and Joseph J. Collins. *Peterson Field Guide to Reptiles and Amphibians of Eastern and Central North America.* Boston: Houghton Mifflin, 1991.

Dickerson, Mary Cynthia. *The Frog Book: North American Toads and Frogs, with a Study of the Habits and Life Histories of Those of the Northeastern States.* New York: Dover, 1969.

Ditmars, Raymond Lee. *Reptiles of the World: Crocodiles, Lizards, Snakes, Turtles and Tortoises of the Eastern and Western Hemispheres.* New York: Macmillan, 1933.

Doubilet, David, "Australia's Saltwater Crocodiles." *National Geographic* (June 1996).

Ernst, Carl, and Roger Barbour. *Turtles of the United States and Canada.* Washington, D.C.: Smithsonian Institution, 2000.

Gans, Carl. *Reptiles of the World.* New York, Bantam, 1975.

Glasgow, Vaughn L. *A Social History of the Alligator.* New York: St. Martin's, 1991.

Guggisberg, Charles Albert Walter. *Crocodiles: Natural History, Folklore, and Conservation.* Mechanicsburg, Pa.: Stackpole, 1972.

Halliday, Tim. *Encyclopedia of Reptiles and Amphibians.* New York: Facts on File, 1986.

Keeton, William T. *Biological Science,* 6th. ed. New York: Norton, 1997.

Kluge, Arnold G., et al. *Chordate Structure and Function,* 2nd ed. New York: Macmillan, 1977.

Leviton, Alan E. *Reptiles and Amphibians of North America.* Garden City, N.Y.: Doubleday, 1971.

Mattison, Christopher. *The Care of Reptiles and Amphibians in Captivity,* rev. ed. New York: Blandford, 1987.

———. *Lizards of the World.* New York: Sterling, 1998.

———. *Snakes of the World.* New York: Sterling, 1998.

———. *Frogs and Toads of the World.* New York: Sterling, 1998.

Obst, Fritz Jürgen. *Turtles, Tortoises and Terrapins.* New York: St. Martin's, 1986.

Pinny, Roy. *The Snake Book.* Garden City, N.Y.: Doubleday, 1981.

Pope, Clifford Hillhouse. *The Giant Snakes: The Natural History of the Boa Constrictor, the Anaconda, and the Largest Pythons.* New York: Knopf, 1961.

Ross, Charles A., ed. *Crocodiles and Alligators.* New York: Facts on File, 1989.

Schmidt, Karl Patterson, and Robert F. Inger. *Living Reptiles of the World.* Garden City, N.Y.: Doubleday, 1966.

Stebbins, Robert Cyril. *Peterson Field Guide to Reptiles and Amphibians of Western North America: Field Marks of All Species in Western North America, Including Baja, California.* Boston: Houghton Mifflin, 1998.

Stidworthy, John. *Snakes of the World.* London: Hamlyn, 1982.

Birds

Austin, Oliver Luther. *Families of Birds.* New York: Golden, 1971.

———. *Birds of the World: A Survey of the Twenty-Seven Orders and One Hundred and Fifty-Five Families.* New York: Golden, 1965.

Beebe, Frank Lyman, and Harold Melvin Webster. *North American Falconry and Hunting Hawks.* Denver, Colo.: North American Falconry and Hunting Hawks, 1964.

Bull, John, and John Farrand Jr. *Audubon Society Field Guide to North American Birds: Eastern Region,* rev. 2nd ed. New York: Knopf, 1994.

Burton, Maurice, ed. *The New Larousse Encyclopedia of Animal Life.* New York: Bonanza, 1981.

Burton, Philip. *Birds of Prey.* New York: Gallery/Smith, 1989.

Clark, William S. *Peterson Field Guide to Hawks of North America.* Boston: Houghton Mifflin, 1998.

CNN News Broadcast, October 10, 1991.

Connor, Jack. *The Complete Birder: A Guide to Better Birding.* Boston: Houghton Mifflin, 1988.

DeGraff, J., et al. *Forest and Rangeland Birds of the United States: Natural History and Habitat Use.* Washington, D.C.: United States Department of Agriculture, Forest Service, 1991.

Dolesh, Richard J. "Lords of the Shallows: The Great Blue Heron." *National Geographic* (April 1984).

Dunne, Pete. *Hawks in Flight.* Boston: Houghton Mifflin, 1988.

Elman, Robert. *The Hunter's Field Guide to the Game Birds and Animals of North America.* New York: Knopf, 1982.

Farrand, John. *Audubon Handbook: How to Identify Birds.* New York: McGraw Hill, 1988.

———. *Audubon Handbook: Eastern Birds.* New York: McGraw Hill, 1988.

———, ed. *Audubon Master Guide to Birds,* Vol. 2. New York: Knopf, 1983.

"Fixit." *Star Tribune* (Minneapolis), March 12, 1989.

Glasier, Phillip. *Falconry and Hawking,* 3rd ed. London: Overlook, 1998.

Goodwin, Derek. *Crows of the World,* 2nd ed. Seattle: University of Washington Press, 1986.

———. *Pigeons and Doves of the World.* London: British Museum Press, 1977.

Hancock, James. *The Herons Handbook.* North Pomfret, Vt.: Trafalgar Square, 1990.

Harper, Don. *The Practical Encyclopedia of Pet Birds for Home and Garden.* New York: Harmony, 1987.

Harrison, Peter, and Tory Harrison. *Sea Birds: An Identification Guide.* Boston: Houghton Mifflin, 1991.

Hosking, Eric John. *Birds of Prey of the World.* New York: Viking, Penguin, 1987.

Johnsgard, Paul A. *Hawks, Eagles, and Falcons of North America: Biology and Natural History.* Washington, D.C.: Smithsonian Institution, 1990.

———. *Grouse and Quails of North America.* Lincoln: University of Nebraska Press, 1973.

Kilham, Lawrence. *The American Crow and the Common Raven.* College Station: Texas A & M University Press, 1991.

Martin, Richard Mark. *Encyclopedia of Aviculture: Keeping and Breeding Birds.* New York: Arco, 1983.

National Geographic Society Field Guide to the Birds of North America, 2nd ed. Washington, D.C.: National Geographic Society, 1987.

Newton, Ian, ed. *Birds of Prey.* New York: Facts on File, 1990.

Orians, Gorden H. *Blackbirds of the Americas.* Seattle: University of Washington Press, 1985.

Peterson, Roger Tory. *Peterson Field Guide to Western Birds: A Completely New Guide to Field Marks of All Species Found in North America West of the 100th Meridian,* 3rd ed. Boston: Houghton Mifflin, 1990.

Ransom, Jay Ellis, comp. *Harper & Row Complete Field Guide to North American Wildlife.* New York: Harper and Row, 1981.

Rowley, Sam R. *Discovering Falconry: A Comprehensive Guide to Contemporary Falconry.* Dexter, N.Y.: New Dawn, 1985.

Samson, Jack. *Modern Falconry: Your Illustrated Guide to the Art and Sport of Hunting with North American Hawks.* Mechanicsburg, Pa.: Stackpole Books, 1984.

Scott, Shirley, ed. *National Geographic Field Guide to Birds of North America,* 2nd ed. Washington, D.C.: National Geographic Society, 1987.

Terres, John K. *The Audubon Society Encyclopedia of North American Birds.* New York: Wings, 1991.

Weidensaul, Scott. *Raptors: The Birds of Prey.* New York: Lyons and Burford, 1996.

Young, Steven. *To the Arctic: An Introduction to the Far Northern World.* New York: Wiley, 1989.

Mammals

Alderton, David. *Foxes, Wolves and Wild Dogs of the World.* New York: Facts on File, 1994.

Allen, Thomas B., ed. *Wild Animals of North America.* Washington, D.C.: National Geographic Society, 1995.

Ashworth, William. *Bears, Their Life and Behavior: A Photographic Study of North American Species.* New York: Crown, 1992.

Ayto, John. *Dictionary of Word Origins.* New York: Little, Brown, 1991.

Baker, Mary L. *Whales, Dolphins, and Porpoises of the World.* Garden City, N.Y.: Doubleday, 1987.

Bateman, Graham, project ed. *All the World's Animals: Sea Mammals.* New York: Torstar, 1984.

Bauer, Erwin A. *Erwin Bauer's Predators of North America.* New York: Outdoor Life, 1988.

———. *Bears.* Stillwater, Minn.: Voyager, 1996.

Bisacve, Michael, et al, eds. *Illustrated Encyclopedia of Animals.* New York: Exeter, 1984.

Brown, Gary. *The Great Bear Almanac.* New York: Lyons and Burford, 1993.

Burt, William Henry, and Richard P. Grossenheider. *A Field Guide to the Mammals: North America North of Mexico,* 3rd ed. Peterson Field Guide Series. Boston: Houghton Mifflin, 1998.

Burton, Maurice, ed. *The New Larousse Encyclopedia of Animal Life.* New York: Bonanza, 1981.

Busch, Robert H. *The Wolf Almanac.* New York: Lyons and Burford, 1995.

Clutton-Brock, Juliet. *Horsepower: A History of the Horse and the Donkey in Human Societies.* Cambridge, Mass.: Harvard University Press, 1992.

———. *Domesticated Animals from Early Times.* Austin: University of Texas Press, 1981.

Coffey, David J. *Dolphins, Whales, and Porpoises: An Encyclopedia of Sea Mammals.* New York: Macmillan, 1977.

Considine, Douglas M., ed. *Van Nostrand's Scientific Encyclopedia,* 8th ed. New York: Van Nostrand Reinhold, 1995.

Craighead, Frank C. *Track of the Grizzly.* San Francisco: Sierra Club, 1982.

Cramond, Mike. *Of Bears and Man.* Norman: University of Oklahoma Press, 1986.

Domico, Terry. *Kangaroos: The Marvelous Mob.* New York: Facts on File, 1993.

Doubilet, David. "Australia's Saltwater Crocodiles." *National Geographic* (June 1996).

Ellis, Richard. *Dolphins and Porpoises.* New York: Knopf, 1989.

———. *The Book of Whales.* New York: Knopf, 1985.

Elman, Robert. *The Hunter's Field Guide to the Game Birds and Animals of North America.* New York: Knopf, 1987.

Ensminger, M. E. *Animal Science,* 9th ed. Danville, Ill.: Interstate, 1991.

———. *The Stockman's Handbook,* 7th ed. Danville, Ill.: Interstate, 1992.

Estes, Richard D. *The Safari Companion: A Guide to Watching African Mammals.* Post Mills, Vt.: Chelsea Green, 1993.

———. *The Behavior Guide to African Mammals.* Berkeley: University of California Press, 1991.

Evans, Peter G. H. *The Natural History of Whales and Dolphins.* New York: Facts on File, 1987.

Feltwell, John. *The Encyclopedia of Butterflies.* New York: Prentice Hall, 1993.

Fox, Michael W. *The Dog: Its Domestication and Behavior.* New York: Garland STPM, 1978.

Gotch, A. F. *Latin Names Explained: A Guide to the Scientific Classification of Reptiles, Birds and Mammals.* New York: Facts on File, 1995.

Groves, Donald G. *The Ocean World Encyclopedia.* New York: McGraw-Hill, 1980.

Hall, Raymond E. *Mammals of the World,* 2nd ed. New York: Wiley, 1981.

Heiser Jr., Charles B. *Seed to Civilization: The Story of Food,* 3rd ed. Cambridge, Mass.: Harvard University Press, 1990.

Hutchins, Betsy, and Paul Hutchins. *The Definitive Donkey: A Textbook on the Modern Ass.* Denton, Tex.: American Donkey and Mule Society, 1999.

Hyams, Edward. *Animals in the Service of Man.* Philadelphia: Lippincott, 1972.

Iljin, N. A. "Wolf-Dog Genetics." *Journal of Genetics,* no. 42 (1941): 359–414.

Jones Jr., J. Knox, David M. Armstrong, and Jerry R. Choate. *Guide to Mammals of the Plains States.* Lincoln: University of Nebraska Press, 1985.

Kid, Jane. *The International Encyclopedia of Horses and Ponies.* New York: Macmillan, 1995.

King, Judith E. *Seals of the World,* 2nd ed. Ithaca, N.Y.: Comstock, 1983.

Lawrence, Eleanor. *Henderson's Dictionary of Biological Terms.* New York: Wiley, 1989.

Linehan, Edward J. "The Trouble with Dolphins." *National Geographic* (April 1979).

Lopez, Barry Holstun. *Of Wolves and Men.* New York: Scribner's, 1982.

Mannucci, Maria Pia, and Alessandro Minelli. *Great Book of the Animal Kingdom.* New York: Arch Cape, 1988.

McGrayne, Sharon Bertsch. *365 Surprising Scientific Facts, Breakthroughs, and Discoveries.* New York: Wiley, 1994.

McLoughlin, John C. *The Canine Clan: A New Look at Man's Best Friend.* New York: Viking, 1983.

Mech, David L. *The Wolf: The Ecology and Behavior of an Endangered Species.* Minneapolis: University of Minnesota Press, 1985.

Miller, Stephan A., and John P. Harley. *Zoology: The Animal Kingdom.* Dubuque, Iowa: Brown, 1996.

Mitchell, Jack Douglas. *1992 Remington Calendar.* Wilmington, Del.: Remington Arms, 1992.

Murry, John A. *Grizzly Bears: An Illustrated Field Guide.* Boulder, Colo.: Roberts Rinehart, 1995.

Norman, David. *Prehistoric Life: The Rise of the Vertebrates.* New York: Macmillan, 1994.

Nowak, Ronald M. *Walker's Mammals of the World,* 6th ed. Baltimore: Johns Hopkins University Press, 1999.

Parker, Gerry. *Eastern Coyote: The Story of Its Success.* Halifax, Nova Scotia: Nimbus, 1995.

Rennicke, Jeff. *Bears of Alaska in Life and Legend.* Boulder, Colo.: Roberts Rinehart, 1987.

Reynolds III, John E., and Daniel K. Odell. *Manatees and Dugongs.* New York: Facts on File, 1991.

Ryden, Hope. *God's Dog: A Celebration of the North American Coyote.* New York: Lyons and Burford, 1997.

Sanderson, Ivan Terence. *Living Mammals of the World.* Garden City, N.Y.: Doubleday, 1995.

Savage, R. J. G., and M. R. Long. *Mammal Evolution: An Illustrated Guide.* New York: Facts on File, 1986.

Senate Select Committee on Animal Welfare. *Kangaroos: Report by the Senate Select Committee on Animal Welfare.* Canberra, Australia: Commonwealth of Australia, 1988.

Shipman, Paul. "Evolution Watch: All in the Family." *Discover* (July 1989).

Shoshani, Jeheskel, cons. ed. *Elephants: Majestic Creatures of the Wild.* Emmaus, Pa.: Rodale, 1992.

Smythe, R. H. *The Dog: Structure and Movement.* New York: Arco, 1970.

Stirling, Ian, cons. ed. *Bears: Majestic Creatures of the Wild.* Emmaus, Pa.: Rodale, 1993.

Turbak, Gary. *Twilight Hunters: Wolves, Coyotes, and Foxes.* Flagstaff, Ariz.: Northland, 1987.

Whitfield, Philip, ed. *Macmillan Illustrated Animal Encyclopedia.* New York: Macmillan, 1999.

Wild Animals of North America, rev. Washington, D.C.: National Geographic Society, 1995.

Young, J. Z. *The Life of Vertebrates,* 3rd ed. New York: Oxford University Press, 1991.

Index